Neuropeptides and Other Bioactive Peptides: From Discovery to Function

Colloquium Series on Neuropeptides

Communication between cells is essential in all multicellular organisms, and even in many unicellular organisms. A variety of molecules are used for cell-cell signaling, including small molecules, proteins, and peptides. The term 'neuropeptide' refers specifically to peptides that function as neurotransmitters, and includes some peptides that also function in the endocrine system as peptide hormones. Neuropeptides represent the largest group of neurotransmitters, with hundreds of biologically active peptides and dozens of neuropeptide receptors known in mammalian systems, and many more peptides and receptors identified in invertebrate systems. In addition, a large number of peptides have been identified but not yet characterized in terms of function. The known functions of neuropeptides include a variety of physiological and behavioral processes such as feeding and body weight regulation, reproduction, anxiety, depression, pain, reward pathways, social behavior, and memory. This series will present the various neuropeptide systems and other aspects of neuropeptides (such as peptide biosynthesis), with individual volumes contributed by experts in the field.

Published titles

(To see published titles please go to the website, www.morganclaypool.com/page/lifesci)

Neuropeptides and Other Bioactive Peptides: From Discovery to Function
Lloyd D. Fricker
www.morganclaypool.com

ISBN: 9781615045068 paperback

ISBN: 9781615045075 ebook

DOI: 10.4199/C00058ED1V01Y201205NPE003

A Publication in the

COLLOQUIUM SERIES ON NEUROPEPTIDES

Lecture #3

Series Editors: Lloyd D. Fricker, Albert Einstein College of Medicine

Series ISSN Pending
ISSN 2154-560X print
ISSN 2154-5626 electronic

Neuropeptides and Other Bioactive Peptides: From Discovery to Function

Lloyd D. Fricker
Albert Einstein College of Medicine

COLLOQUIUM SERIES ON NEUROPEPTIDES #3

ABSTRACT

Neuropeptides and peptide hormones represent the largest class of chemical messengers that transmit information from one cell to another. In this review, several decades of research on peptides in cell-cell signaling are summarized, with a focus on neuropeptide discovery, biosynthesis, and function. In addition to covering the well-studied aspects of neuropeptides, emerging concepts are discussed, including classical versus non-classical neuropeptides and direct versus indirect neuropeptides. Other potential functions for peptides in intercellular and intracellular signaling are also discussed.

KEYWORDS

neuropeptide, hormone, prohormone convertase, proprotein convertase, carboxypeptidase, amidation, enkephalin, dynorphin, endorphin, SAAS, neuropeptide Y, hemopressin

Acknowledgments

Thanks to Lakshmi Devi, Iris Lindberg, Emer Ferro, Antonio Camargo, and Matthew Sapio for critically reading a draft of this review and providing helpful comments.

Acknowledgments

[illegible]

Contents

CHAPTER 1

Overview of Neuropeptides

Neuropeptides play important roles in a number of diverse physiological processes, including reproduction, feeding and body weight regulation, pain, memory, mood, anxiety, reward pathways, arousal, and sleep/wake cycles [1]. Neuropeptides are the largest group of cell-cell signaling molecules, with over 100 different peptides known to function in this capacity. In addition, hundreds more peptides have been physically identified but do not have known functions, and some of these peptides are likely to have additional roles in cell-cell communication.

1.1 NEUROPEPTIDES, PEPTIDE HORMONES, AND BIOLOGICALLY ACTIVE PEPTIDE: WHAT IS A PEPTIDE?

Neuropeptides are an important group of "biologically active peptides," but the two terms are not interchangeable; peptides play many important biological roles and only some of these are neuropeptides. The term "neuropeptide" refers to a peptide that conveys information from one cell to another. Other types of biologically active peptides that are not neuropeptides include those that function as antibiotics in humans and many other organisms and peptide toxins such as those present in snakes, spiders, and other species [2–4]. In addition, some peptides found in the brain and/or other tissues of vertebrate animals have functions that are not involved in cell-cell signaling, and therefore, it would be incorrect to refer to these molecules as neuropeptides. For example, glutathione is a small peptide, consisting of 3 amino acids, that is found in the cytosol of all cells, including neurons, where it functions in the reduction of oxidized molecules. There are also small proteins that have clear biological roles that are distinct from the roles of neuropeptides; examples include thymosin beta-4 and thymosin beta-10 [5]. Some people define a peptide as any polymer of amino acids that is smaller than 10 kDa, and this definition includes the thymosins and other small proteins. Others define a peptide as a fragment of a protein, which would not only exclude the thymosins and small cellular proteins but also glutathione and other cytosolic peptides made directly from amino acids by a process that does not involve ribosomes. In this review, the term "peptide" refers to any polymer of amino acids between 2 and 100 amino acids, excluding intact proteins.

Neuropeptides have many similarities to peptide hormones, and some of the same peptides are used in both the brain as neuropeptides and the endocrine system as peptide hormones. In both cases, the peptide is secreted from one cell and then signals another cell. The term "neuropeptide" was created as a contraction of "neurotransmitter peptide" and was originally applied only to peptides secreted from neurons. However, some neuropeptides are produced in glial cells as well as neurons, and so the term "neuropeptide" is now used to refer to peptides that function in cell-cell communication in the brain, while the term "peptide hormone" is used to refer to those peptides that function in the periphery to signal between organs. Collectively, neuropeptides and peptide hormones are occasionally referred to as neuroendocrine peptides, although this term does not imply function as a cell-cell signaling molecule but rather that the peptide is found in the neuroendocrine system, much like "brain peptide" refers to all peptides found in the brain. To qualify as a neuropeptide (or peptide hormone), the peptide needs to meet several criteria that are based on those originally developed to classify neurotransmitters and hormones (described below). Neuropeptides and peptide hormones are also referred to as "regulatory peptides," which includes peptides of the gut, endocrine, and nervous systems, which regulate cell or tissue function.

1.2 SIMILARITIES BETWEEN CLASSICAL NEUROTRANSMITTERS AND NEUROPEPTIDES

The classical definition of cell-cell signaling molecules includes well-known neurotransmitters such as acetylcholine, dopamine, and serotonin and also neuropeptides. All of these molecules are produced in neurons and secreted in a regulated fashion (Text Box 1). Once secreted, the molecule binds to a receptor protein that influences cellular activity. Finally, there are mechanisms for the removal of the molecule from the extracellular space. These criteria were developed for the first cell-cell signaling molecules and are applicable for most of the cell-cell signaling molecules subsequently

Box 1. Classical Definition of Cell–Cell Signaling Molecules (Neurotransmitters and Neuropeptides)

Molecule is produced in cell
Molecule is secreted from cell
Secretion is regulated
Molecule binds to receptor on another cell, thereby affecting cellular activity
Molecule is removed from the extracellular space by metabolism or reuptake

discovered, including peptides. However, some recently discovered peptides do not fit these criteria and have been named "non-classical neuropeptides"; these are described in the next section. Unless specified, the term "neuropeptide" is used to refer to classical neuropeptides that meet the criteria for classical cell-cell signaling molecules.

Classical neuropeptides are synthesized within cells from specific precursors; these precursors generally do not have functions on their own, requiring processing to generate the active forms. The precursors are cleaved into smaller forms by a series of enzymes that act on specific cleavage sites to first generate processing intermediates and then after further cleavages and/or other post-translational modifications, the mature forms of the peptides (Figure 1). These cleavages and other modifications occur within the secretory pathway. The neuropeptide precursor is commonly referred to as a prohormone, based on similarity in the biosynthetic pathway of peptide hormones and neuropeptides. Although it would make sense to call the neuropeptide precursor a "proneuropeptide," this term is rarely used. Prohormones are transported into the lumen of the endoplasmic reticulum during protein synthesis due to the presence of an N-terminal sequence dubbed the "signal peptide." This signal peptide exists only transiently when it is cleaved from the nascent protein by a signal peptidase, and soon after cleavage the peptide is degraded by a peptidase present in membranes of the endoplasmic reticulum.

Once in the secretory pathway, the prohormone may be modified by enzymes in the endoplasmic reticulum and Golgi apparatus, which add carbohydrates, phosphates, and/or sulfates. In the late part of the Golgi, termed the trans-Golgi network, the prohormone is sorted into the regulated secretory pathway and then processed into smaller peptides (Figure 1). In some cases, processing of the prohormone begins prior to sorting into the regulated secretory vesicles, while for others, the processing does not start until after the sorting is complete. Most peptide precursors are processed by an endopeptidase such as prohormone convertase 1 (PC1, also known as PC3, and often referred to as PC1/3) and prohormone convertase 2 (PC2), which generate intermediates containing C-terminal basic residues that then are removed by carboxypeptidase D (CPD) or carboxypeptidase E (CPE). In some cases, additional modifications occur. The final mature (and usually bioactive) form of the peptide is stored within secretory vesicles (also called secretory granules). These granules accumulate within the cell. Stimulation of the cell, typically by depolarization, causes the vesicles to fuse with the cellular membrane, thereby releasing the contents into the extracellular environment. It is this regulated secretion that results in the large increase in extracellular levels of the peptide. The production of peptides can be regulated at several levels; gene transcription, mRNA stability, translation, and processing; these lead to altered amounts or forms of peptide in the vesicles, but ultimately, it is the regulation of secretion that is most important in controlling extracellular levels of the classical neuropeptides.

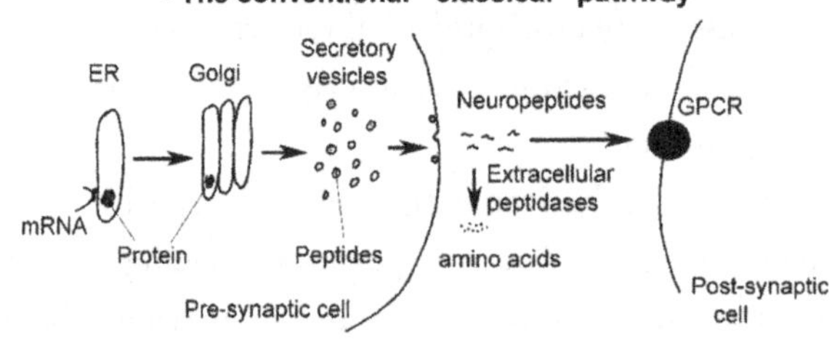

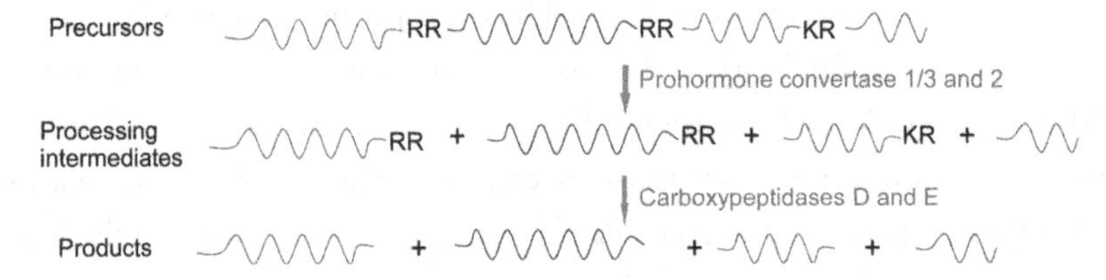

FIGURE 1: Overview of peptide biosynthesis. Top: In the classical pathway, neuropeptides, and peptide hormones are initially produced as precursor proteins in the endoplasmic reticulum (ER) as cytosolic mRNA is translated into protein and the growing protein is translocated into the lumen of the ER. After folding of the protein and addition of N-linked carbohydrate (if the appropriate sites are present in the protein), the protein is transported to the Golgi via small vesicles that bud from the ER and fuse with the Golgi. In the latter part of the Golgi, termed the trans-Golgi network, the peptide precursor is sorted into immature secretory granules, together with peptide processing peptidases that proceed to cleave the precursor. Upon stimuli, the secretory granules fuse with the plasma membrane to release their cargo into the extracellular environment. The secreted peptide can either bind to a receptor or get processed into smaller peptides (not shown) or amino acids. The receptors are often G-protein-coupled receptors (GPCRs). Bottom: Initial cleavages of the peptide precursor are mediated by endopeptidases (prohormone convertases 1/3 and 2); these enzymes convert the precursor into intermediates containing C-terminal basic amino acids: Lys (K) and Arg (R). The intermediates are converted into the mature peptides primarily by carboxypeptidase E, although carboxypeptidase D also participates in this reaction. Some peptides require additional post-translational modifications such as C-terminal amidation (not shown).

Upon release from the cell, the peptide diffuses throughout the extracellular environment where it can interact with receptors containing specific binding sites (Figure 1). The interaction of the peptide with the receptor causes a conformational change in the receptor, and this change produces a cellular response through a variety of mechanisms depending on the type of receptor and the second messenger systems present in the cell. Signaling molecules need to have a mechanism for the termination of action so that the molecule produces an effect only for a short time. Initially, it was thought that there were unique degradative enzymes for each neuropeptide, raising the hopes that this system would be amenable to drug development [6]. However, further studies found that the "specific" enzymes were able to cleave many different neuropeptides, and the current consensus in the field is that a relatively small number of extracellular proteases/peptidases are capable of cleaving a variety of neuropeptides and peptide hormones.

1.3 DIFFERENCES BETWEEN CLASSICAL NEUROTRANSMITTERS AND NEUROPEPTIDES

The above criteria apply to both classical neurotransmitters and neuropeptides although there are important differences between these two groups. In general, classical neurotransmitters are produced in the synaptic terminus of the neuron and transported into vesicles that are secreted from specialized synaptic areas upon neuronal stimulation (Figure 2). Most extracellular neurotransmitter can be reinternalized into the presynaptic termini via specific transporters; an exception is acetylcholine, which is first broken into acetate and choline, which are then transported into the presynaptic termini and converted into acetylcholine. In contrast, neuropeptides are produced from proteins that are translated into the endoplasmic reticulum, transported by a vesicle-mediated process to the Golgi, and then sorted into immature secretory granules. Processing of the prohormone into the bioactive peptides usually requires a series of enzymes and generally occurs as the immature vesicle is transported from the Golgi to the cell surface. Peptide-containing secretory granules are named large dense-core granules. The different classes of vesicles are released by different types of stimulation; the small secretory vesicles require only low frequency stimulation for their release, whereas peptide-containing secretory granules require higher-frequency stimulation before they are released [7]. Also, peptide-containing secretory granules can be released from many parts of a neuron, in contrast to the release of neurotransmitters from small synaptic vesicles, which is limited to presynaptic terminals [7]. Some peptide-containing secretory granules also contain classical neurotransmitters, allowing for diverse signals to be released at the same time.

Neuropeptides are not known to be reinternalized after secretion and are instead degraded by extracellular peptidases, eventually into amino acids that are reabsorbed into neurons and other brain cells. This property of neuropeptides is more similar to acetylcholine than to the other classical neurotransmitters, which are reinternalized without any processing (such as dopamine). But unlike acetylcholine, which is either acetylcholine (which is biologically active) or acetate and choline (which are not active as neurotransmitters), the extracellular processing of peptides does not always lead to their immediate inactivation. Instead, peptides can undergo processing that alters their biological activity. A classic example of this is the peptide bradykinin, which is released from its precursor (kininogen) by the enzyme kallikrein and further processed a carboxypeptidase, which

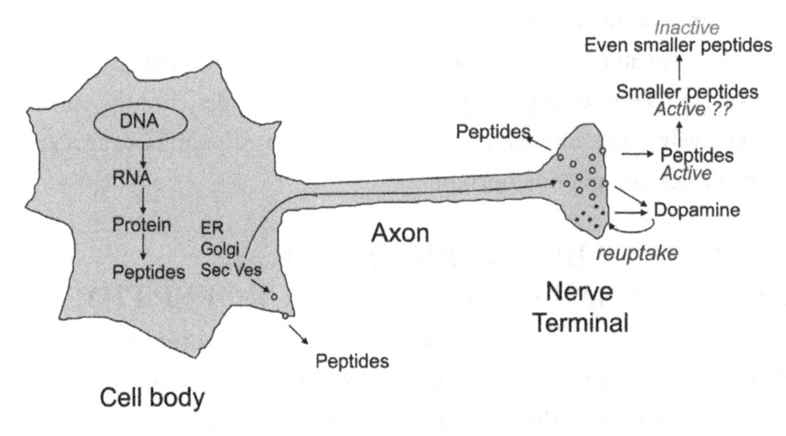

FIGURE 2: Comparison of classical neurotransmitters and neuropeptides. As described in more detail in Figure 1, neuropeptides are produced from larger proteins that are imported into the endoplasmic reticulum (ER), pass through the Golgi apparatus, and are cleaved within secretory vesicles that are transported to the axonal terminal where the large dense-core secretory granules are stored until release. Peptides can also be released from non-synaptic sites, including the cell body and dendrites. In contrast, classical neurotransmitters such as dopamine are produced from enzymes located in the presynaptic nerve terminal and stored in small synaptic vesicles (and on occasion, co-stored with neuropeptides in large dense-core granules). The small synaptic vesicles are released from presynaptic terminals, and the released dopamine is reabsorbed into the nerve terminal by selective transporters. Peptides are not reabsorbed after release and are instead broken down by peptidases located in the extracellular environment. In some cases, the smaller peptides produced from the secreted peptide retain some of the biological activities of the original peptide or have distinct activities, and so the extracellular processing is not always degradative. Eventually the extracellular peptides are converted into inactive products and amino acids, which are reabsorbed (not shown).

removes the C-terminal arginine [8]. The full-length form of bradykinin binds to B2 bradykinin receptors, while the shorter form lacking the C-terminal arginine binds with higher affinity to the B1 bradykinin receptor [9].

The intravesicular processing of peptides is also a point at which the biological activity of the neuropeptide or peptide hormone can be regulated. A classic example of this is the processing of the precursor proopiomelanocortin in the pituitary (Figure 3). In the anterior lobe of the pituitary, the precursor is converted into adrenocorticotropic hormone (ACTH), a 39-residue peptide that binds to several subtypes of melanocortin receptors such as the MC2R located in the adrenal gland; this receptor controls the production of glucocorticoids (cortisol) [10, 11]. In the intermediate lobe

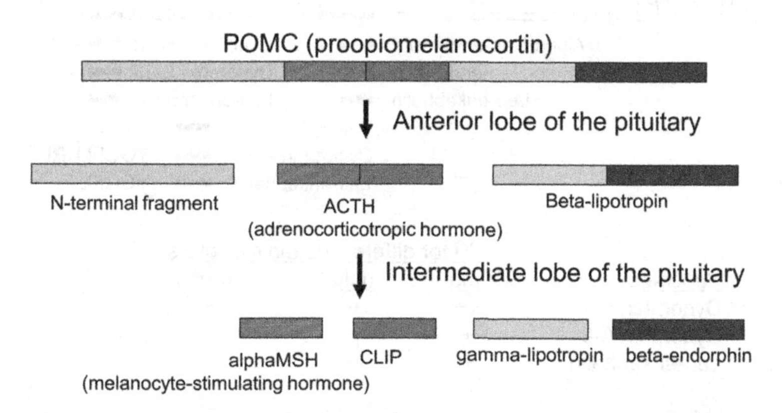

FIGURE 3: Intracellular processing of the prohormone can produce peptides with distinct functions. In this example, proopiomelanocortin (POMC) is processed by prohormone convertases and carboxypeptidases (and for some peptides, additional enzymes that perform post-translational modifications). In the anterior lobe of the pituitary, POMC is processed into adrenocorticotropic hormone (ACTH), which stimulates the adrenal through melanocortin 2 receptors. In the intermediate lobe of the pituitary, ACTH is further processed into alpha-melanocyte-stimulating hormone (α-MSH), which has no affinity for melanocortin 2 receptors but binds to other melanocortin receptors found in the periphery (on melanocytes) and in the brain.

of the pituitary and also in the brain, ACTH is further processed into alpha-melanocyte-stimulating hormone (α-MSH), a 13-residue peptide that binds to melanocortin receptors with completely different affinities than ACTH [11, 12]. There are many other examples of this, some as extreme as that of ACTH and α-MSH, and others where the difference is more subtle. Examples where the processing leads to a gradual effect are found in the prodynorphin system. This precursor contains three major bioactive domains that if completely processed would yield the 5-residue peptide named Leu-enkephalin (Figure 4). However, processing of prodynorphin in the brain does not generally go as far as enkephalin, and larger enkephalin-containing peptides are the major products [13]. The larger products such as dynorphin A-17 bind to the kappa opioid receptor with the highest affinity and are only weak ligands for the delta opioid receptor [14]. Shorter forms, such as dynorphin A-8, are slightly less potent at the kappa receptor, and enkephalin is much less potent [14, 15]. In contrast, enkephalin is more potent at the delta receptor than the longer enkephalin-containing

Differential processing of the precursor can produce peptides with slightly different activities

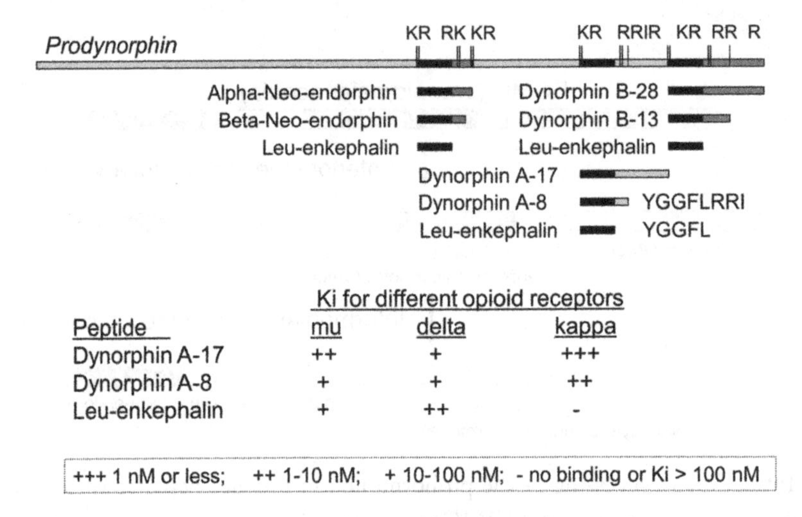

Peptide	Ki for different opioid receptors: mu	delta	kappa
Dynorphin A-17	++	+	+++
Dynorphin A-8	+	+	++
Leu-enkephalin	+	++	-

FIGURE 4: Intracellular processing of the prohormone/neuropeptide precursor can alter the affinity of peptides for related receptors. In this example, prodynorphin is processed into a range of peptides that are able to bind to the mu, delta, and kappa opioid receptors. All of these opioid receptor-binding peptides contain the N-terminal Leu-enkephalin sequence (Tyr-Gly-Gly-Phe-Leu). Those peptides with C-terminal extensions bind to kappa receptors with higher affinity than Leu-enkephalin, while binding to delta receptors shows the opposite trend, with the highest affinity observed with the 5-residue enkephalin peptide.

dynorphins [14, 15]. Thus, processing of the peptide has a dramatic influence on the relative affinity of the resulting peptides for the various receptors.

1.4 NON-CLASSICAL NEUROTRANSMITTERS AND NEUROPEPTIDES

The classical definition of neurotransmitters, which was established many decades ago, does not apply to some types of signaling molecules discovered in the past 10–20 years. For example, nitric oxide and lipid-based endocannabinoids have distinct properties and are often referred to as "non-classical neurotransmitter" or neuromodulators [16, 17]. Unlike classical neurotransmitters, which are synthesized in advance and stored within secretory vesicles until needed, the non-classical neurotransmitters are produced when needed by a regulated process and immediately released from the cell. Both nitric oxide and the lipid-based endocannabinoids are produced from abundant cellular molecules; nitric oxide is made from arginine and the lipid-based endocannabinoids are made from phospholipids. Thus, there is no need for the bioactive molecules to be produced in advance and stored within vesicles until needed—the cell can produce large amounts simply by activating the enzymes that convert the arginine or phospholipids into the bioactive moiety.

Over the past several decades, a number of bioactive peptides have been isolated from the brain and other tissues as a result of searches for molecules that bind to specific receptors (this is described in more detail in Chapter 2). Many of these novel peptides were found to be classical neuropeptides, produced in the secretory pathway, but some of the novel peptides were derived from cytosolic proteins. Many scientists (including this author) initially expressed doubts that these peptides could function in cell-cell signaling due to their incorrect intracellular localization. However, recent studies have found that some of these peptides can be secreted from cultured brain tissue (although the mechanism is not yet known) [18]. In addition, some of the intracellular protein-derived peptides appear to be up-regulated following specific stimuli [19]. Taken together with the ability of the peptide to bind to cell surface receptors and elicit changes, it is possible that these cytosol-derived peptides represent a new class of neuropeptide. By analogy with non-classical neurotransmitters, an appropriate name for this class of peptides is "non-classical neuropeptides" (Figure 5). Specific examples of non-classical neuropeptides are described in Chapter 5.

In this book, the term non-classical neuropeptide will be used to refer to peptides that are produced from intracellular proteins, secreted from the cell, and able to cause an effect on neighboring cells, thus playing a role in cell-cell signaling. The criteria for non-classical cell-cell signaling molecules, whether neurotransmitter or peptide, are listed in Text Box 2; these criteria are similar to those described for classical cell-cell signaling molecules (Text Box 1), except for two differences. First, the concept of regulated secretion (a criterion for classical neurotransmitters) is loosened to allow for regulated synthesis followed by constitutive secretion. As long as the extracellular levels

Box 2. Broad Definition of Cell–Cell Signaling Molecule (Including Classical and Non-Classical)

- Molecule is produced in cell
- Molecule is secreted from cell
- Extracellular levels of molecule are regulated, either by regulation of secretion or by regulation of production
- Molecule influences another cell either by binding to a receptor or through another process
- Molecule is removed from the extracellular space by metabolism or reuptake

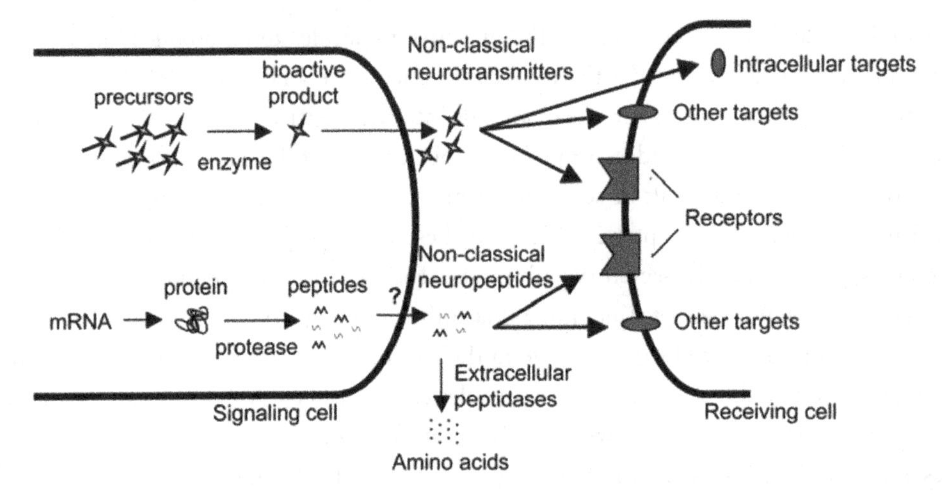

FIGURE 5: Non-classical neurotransmitters and neuropeptides. Non-classical neurotransmitters include nitric oxide (NO) and the lipid-based endocannabinoids; these molecules are produced in the cell cytosol upon the appropriate stimulation and then released. Neighboring cells are affected by these released molecules, either through cell surface receptors (in the case of the endocannabinoids) or intracellular targets (for NO). Likewise, the proposed role for non-classical neuropeptides includes synthesis in the cytosol by proteolytic processing of precursor proteins followed by constitutive release. The mechanism of this release is not yet known. Once secreted, the peptide can influence receptors on nearby cells, either functioning through a cell surface receptor or another target. Figure modified from Gelman and Fricker [20].

of the molecule are altered, it does not matter to the responding cell whether the message was secreted or synthesized via a regulated process. Second, the requirement that the cell-cell signaling molecule interacts with a receptor has been eliminated and replaced with the simple requirement that the substance produce an effect on the responding cell. Not all signaling mechanisms need to go through receptors. For example, the non-classical neurotransmitter nitric oxide has a variety of mechanisms of interacting with neighboring cells: modulation of guanylate cyclase activity, *S*-nitrosylation of cytosolic proteins, and other activities [21, 22]. It is possible that peptides also have non-receptor functions. Some peptides that produce a biological effect on other cells or tissues may not directly bind to receptors but instead exert their effect through inhibition of extracellular peptidases or other mechanisms (described below).

It should be emphasized that the concept of non-classical neuropeptides is neither well established nor widely known. Although many of the parts are in place, and there is a direct analogy between non-classical neurotransmitters and non-classical neuropeptides, it remains to be demonstrated that non-classical peptides can actually function in a physiological system. To establish a molecule as a classical or non-classical neuropeptide, the above criteria need to be demonstrated, but this is not enough—it is also necessary to show that the peptide is required for cell-cell signaling in a biological system by removing the peptide and observing a change in a physiological parameter. Years ago, the major technique to reduce the levels of peptide in the extracellular environment was antibody neutralization. In this technique, antibodies to the peptide are applied to a biological system (such as by injection into mouse brain) and then the system is monitored for a change in behavior or another parameter. This technique is still used, as it is fairly straightforward and direct, but other approaches are now more common, such as gene disruption and RNA interference such as small interfering RNA (siRNA) and short hairpin RNA (shRNA). The advantage of gene disruption is that it completely eliminates the endogenous peptide, which is not possible with antibody neutralization. However, if multiple peptides are encoded by the same gene (which is very common), then this approach will not reveal which of the encoded peptides are responsible for a particular physiological process. Combinations of these techniques to remove or reduce the levels of specific peptides are necessary to establish the precise function of a given neuropeptide. Furthermore, the complication of disrupting genes for cellular proteins that serve additional functions makes it difficult to use these approaches to test the non-classical neuropeptide hypothesis.

1.5 OTHER POTENTIAL ROLES FOR BIOACTIVE PEPTIDES IN CELL SIGNALING

Although the focus of this review is on neuropeptides, which function in cell-cell signaling through interactions with cell surface receptors, there are other emerging roles for peptides in signaling. For example, angiotensin IV and several other secreted peptides have been shown to inhibit an extracellular peptidase that degrades neuropeptides, thereby altering synaptic levels of neuropeptides and producing a biological effect [23, 24]. If the peptide that inhibits the peptidase has no direct effect on neuropeptide receptors, an appropriate term for this type of biologically active peptide is "indirect neuropeptide" (Figure 6B). This term is related to the pharmacological term "indirect agonist," which is used to describe molecules that are not direct agonists but which alter the levels of endogenous receptor agonists to produce an agonist-like response. For example, inhibitors of acetylcholine esterase block the enzyme that degrades acetylcholine, leading to an agonist-like response. These acetylcholine esterase inhibitors are referred to as indirect agonists. Similarly, indirect neuropeptides are not technically neuropeptides in that they do not bind to receptors, but alter levels of neuropeptides. Examples of indirect neuropeptides, and the enzymes responsible for neuropeptide metabolism in the extracellular space, are described in Chapter 4.

Another emerging idea for peptide function involves peptide interactions with a variety of proteins, not just with receptors and peptidases. This concept is analogous to microRNA, which are small oligomers that play important roles in regulating levels of larger oligomers (i.e., mRNAs) in a

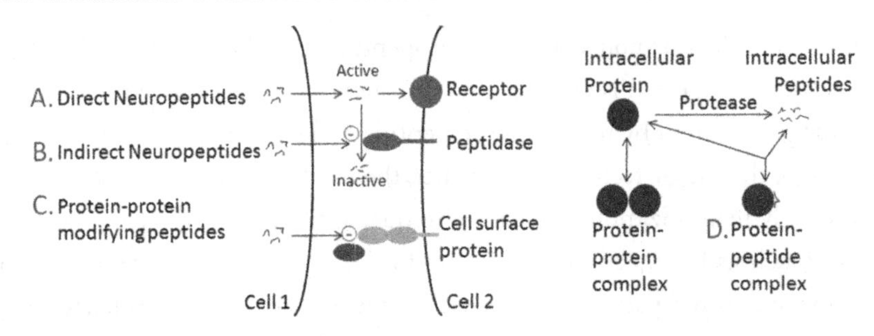

FIGURE 6: Known and theoretical mechanisms for peptides in cell-cell signaling (A–C) and intracellular processes (D). (A) "Direct neuropeptides" includes a large number of known classical neuropeptides (as described in Figure 1) as well as putative non-classical neuropeptides that are derived from cytosolic proteins and secreted by an unconventional pathway (described in Figure 5). (B): "Indirect neuropeptides" refers to peptides that do not directly bind to receptors but which compete for peptidases that degrade neuropeptides. Some evidence exists for peptides that function in this capacity. (C) In theory, secreted peptides may also bind to proteins on the cell surface and either inhibit or mimic protein-protein interactions. (D) In theory, peptides produced in the cytosol or nucleus may bind to proteins and alter their function. Only limited evidence for this has been reported [25], although it is well documented that synthetic peptides of 10–20 amino acids are able to interact with proteins and in many cases affect protein-protein interactions to produce biological effects [25–28].

variety of organisms. For many years, scientists used small RNA molecules to interfere with target mRNAs, without realizing that small RNAs were part of a natural regulatory system. Similarly, many scientists use small peptides of 10–20 amino acids to influence protein-protein interactions within a cell [25–28]. In some cases, peptides bind to proteins and mimic the effect of another protein, and in other cases, the peptide inhibits the protein-protein interaction. Synthetic peptides have also been used to affect protein-substrate interactions or influence protein folding [28, 29]. However, few people consider the possibility that endogenous peptides can influence proteins, largely because of the dogma that peptides have very short half-lives within a cell. This dogma is based on analyses of synthetic peptides modified with fluorescent reporter groups [30, 31]. In contrast to expectation that cytosolic peptides are highly unstable, peptidomic studies examining neuropeptides present in the brain and other biological samples have found a large number of peptides that are derived from cytosolic or mitochondrial proteins [32–34]. The majority of these peptides do not represent the most abundant or unstable cellular proteins, suggesting that they are not simply degradative fragments [34]. Some of the peptides produced from intracellular proteins have been found to be secreted from cells or brain slices and therefore may function as non-classical neuropeptides (which could potentially be either direct or indirect neuropeptides). Other secreted peptides may influence protein-protein interactions between cell surface proteins (Figure 6C). A number of cell surface proteins form protein-protein complexes with proteins in the extracellular matrix or attached to other cells; examples include integrins, neurexins and neuroligins, notch, and many others [35, 36]. While these proteins are commonly referred to as receptors, they do not bind neuropeptides, and their signal transduction mechanisms are distinct from those of the well studied G-protein-coupled receptors. There have not been any reports of secreted peptides influencing the protein-protein interactions of these cell surface proteins, but this remains a theoretical possibility based on the general ability of small peptides to influence protein-protein interactions.

A related idea, also theoretical at present, is that intracellular peptides produce biological effects by influencing protein-protein interactions (Figure 6D). Intracellular peptides are produced during protein degradation, which involves both endopeptidases and exopeptidases. The first step in protein degradation is the action of an endopeptidase that cleaves the protein into peptides. For example, the proteasome is a large multisubunit enzyme complex that cleaves proteins into peptides of 3–22 amino acids [37]. These peptides can be converted into smaller peptides by one or more oligopeptidases present in the cytosol; examples include thimet oligopeptidase (also known as endopeptidase 24.15), neurolysin (also known as endopeptidase 24.16), insulin-degrading enzyme, and prolyl endopeptidase [38–44]. None of these enzymes are efficient at degrading peptides, and all show very limited cleavage in which certain peptides are processed only at specific sites. The initial proteasome product or the products formed by the action of the oligopeptidases are then degraded into amino acids by aminopeptidases in the cytosol [31, 41].

The evidence supporting the function of intracellular signaling peptides is largely circumstantial and is based on the following points. First, many intracellular peptides exist and are detected in mass spectrometry peptidomic studies with signal strengths equivalent to those of well-known neuropeptides; these intracellular peptides are therefore fairly abundant in the brain [32–34]. Second, the proteins that give rise to the observed peptides are not simply the most abundant proteins in the sample nor are they the most unstable proteins [34]. Third, for many proteins, only a small number of peptide fragments are observed in the peptidomics studies, suggesting that these are either selectively generated or selectively retained in the cell, possibly by binding to other proteins [32, 34]. Fourth, when specific peptides have been introduced into cells, they are often able to affect cellular function [25–28]. This is true both for peptides specifically generated to correspond to a protein-protein binding domain and for peptides corresponding to the major peptides detected in peptidomics analyses. While the above points are circumstantial—peptides exist and can be functional—direct proof is needed before this idea can be considered to be an established fact.

• • • •

CHAPTER 2

Neuropeptide Discovery

There are two reasons to consider neuropeptide discovery when learning about peptides. First, there is the historical significance of how we know what we know. But more importantly, by understanding how we have gotten to the current level of knowledge, we are better able to see the gaps in this knowledge and areas for further work. It is extremely likely that additional neuropeptides will be discovered in the future, even in humans and other species that have a sequenced genome. Although it is possible to use the gene sequence to make predictions about the encoded proteins and whether they are likely to be cleaved into peptides and secreted from cells, these predictions are not very accurate and have high rates of both false positives and false negatives. Therefore, there is a need for direct studies to search for novel peptides and identify their functions. It is also likely that many of the peptides already identified from previous studies will be found to have new functions in cell-cell signaling. Ultimately, most biological systems are more complicated than initially thought, and the neuropeptide systems are no exception to this general rule.

To identify a neuropeptide, it is not sufficient to merely learn the sequence of a peptide present in the brain or another tissue—it is also necessary to learn the function(s) of the peptide. Historically, neuropeptides were identified during searches for endogenous molecules that produced a physiological effect. These approaches are referred to as "function first" and require some type of bioassay or other related technique to screen crude tissue extracts (Text Box 3). Once a signal was detected, the tissue was fractionated and each fraction was tested for activity; this process was repeated until the factor was purified to homogeneity and the chemical makeup determined. These function-first approaches are not specific for neuropeptides and have also detected other types of molecules that produce biological effects such as classical neurotransmitters and related molecules. The other general approach to discover neuropeptides is referred to as "peptide first." As the name implies, these techniques start with the identification of peptides and then involve various functional assays to learn which of the identified peptides are biologically active neuropeptides. Some of the peptide-first approaches are aimed at the discovery of classical neuropeptides that are produced in the regulated secretory pathway, while others are broad methods that will find all peptides present in the brain, many of which may not be functional as neuropeptides (but which may perform other functions such as intracellular signaling, described briefly in Chapter 1). For all of these approaches,

Box 3. Approaches to Classical Neuropeptide Discovery

- "Function first" approaches
  - Bioassay (1902 to present)
  - Binding to receptor (1970s to present)
  - Genetic analysis (1990s)
- "Peptide first" approaches
  - Physical properties—C-terminal amide (1980s)
  - cDNA cloning of precursor (1980s to present)
  - Substrates for peptide processing enzymes (2000 to present)
  - Peptidomics of the secretory pathway (2005 to present)

the initial identification is only the first step, and before the peptide can be considered to be a true endogenous neuropeptide, it is necessary for many additional experiments to establish that the peptide meets the criteria discussed in the previous chapter.

2.1 NEUROPEPTIDE DISCOVERY USING BIOASSAYS

The first cell-cell signaling peptide was detected by Bayliss and Starling [45] over a century ago using an assay involving stimulation of acid secretion from the stomach. The factor they found was later identified as secretin once peptide sequencing techniques were developed many decades later [46]. Other peptide hormones were found using different bioassays. For example, in 1909, Henry Dale discovered that an extract from the human posterior pituitary gland contracted the uterus of a pregnant cat, and he named the substance oxytocin. The substance was later found to be a peptide and the sequence was determined in the 1950s by Vincent du Vigneaud [47], who subsequently synthesized the compound and showed that it mimicked the properties of the endogenous hormone. This represented the first synthesis of any bioactive peptide and resulted in a Nobel Prize for du Vigneaud in 1955. As later found for a number of peptide hormones, oxytocin is produced and secreted from neurons, functioning as a neuropeptide in addition to its hormonal role.

The first peptide discovered as a neuropeptide, and not as a hormone, is the 11-residue peptide named substance P. This peptide was discovered in 1931 by Ulf von Euler and John Gaddum, working in the laboratory of Henry Dale [48]. Euler and Gaddum found that an extract of intestine stimulated contraction of the intestinal smooth muscles and also lowered blood pressure but was distinct from acetylcholine because the effect of the unknown substance had a slower onset of action than acetylcholine and was not blocked by atropine (an acetylcholine antagonist). The substance was also found in the brain, and named substance P (the "P" appears to stand for the powder formed during a precipitation step used to partially purify the novel substance away from other molecules known at the time to stimulate smooth muscle contraction). The sequence of substance P was determined 40 years later, in 1971 [49].

Other peptides identified in the 1970s include Met- and Leu-enkephalin. John Hughes, Hans Kosterlitz, and colleagues were interested in the effects of opioid drugs such as morphine, which was known to block the contraction of the mouse vas deferens when electrical stimulation was applied to the tissue. Using this assay, Hughes and Kosterlitz tested brain extracts and found that the extracts produced an effect similar to the opiate compounds (Figure 7). The extracts were fractionated by standard chromatography techniques and then retested; the fractions with biological activity were further purified using additional separation methods until the enkephalins were pure and sequence analysis revealed their identity [50]. Two peptides were identified that differed in one amino acid; these peptides were named Met-enkephalin (Tyr-Gly-Gly-Phe-Met) and Leu-enkephalin (Tyr-Gly-Gly-Phe-Leu). Several years later, Avram Goldstein and colleagues identified

Neuropeptide Discovery: The Bioassay Method

Example: discovery of Met- and Leu-enkephalin

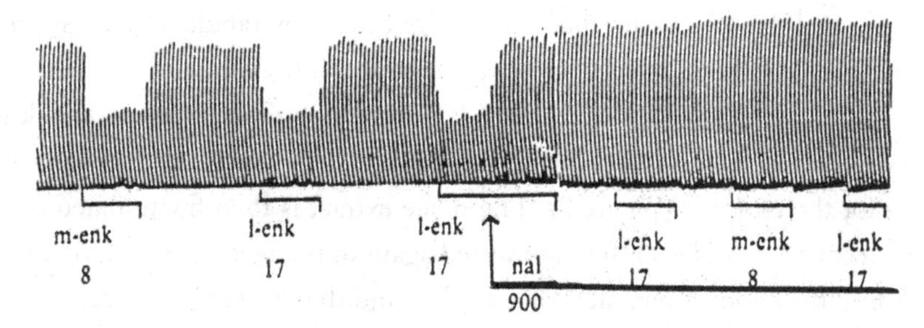

FIGURE 7: Discovery of the enkephalin peptides. Mouse vas deferens was electrically stimulated to induce a contraction (upward spike in tracing). The addition of opioid compounds such as morphine was previously shown to reduce the magnitude of the contractions, and the effect of morphine was blocked by the addition of naloxone. Brain extracts were tested and found to produce a morphine-like response. After purification, the material was found to contain two peptides, Met- and Leu-enkephalin, which were synthesized and tested in the bioassay. Both peptides produced a morphine-like response blocked by administration of naloxone. Figure adapted from Hughes et al. [50].

a peptide from pig pituitary that produced an opioid-like effect on guinea pig ileum; sequencing of the first 13 residues of the novel peptide revealed that it contained the N-terminal five amino acids of Leu-enkephalin followed by a C-terminal extension of unknown length [51]. This peptide was named dynorphin (1–13), and subsequently found to represent the N-terminus of a 17-residue peptide (dynorphin A-17).

Due to the difficulty in finding appropriate bioassays as well as the slow speed of the process, relatively few mammalian neuropeptides have been discovered using the bioassay approach. However, related approaches involving receptor assays have led to many more neuropeptides.

2.2 NEUROPEPTIDE DISCOVERY USING RECEPTOR-BINDING APPROACHES

The bioassay approach described in the previous section does not require knowledge of the specific neuropeptide receptor system involved in the process. A variant of the bioassay approach uses assays based on neuropeptide receptors, which do require knowledge of the receptor systems.

Soon after the development of receptor-binding assays, scientists began to use them to identify the endogenous compounds that activated these receptors. For example, the opioid receptor-binding assay was developed in the early 1970s and then used to rediscover the enkephalin peptides (which had already been discovered using the bioassay method shown in Figure 7) as well as numerous other novel peptides that bound to various opioid receptors: dynorphin B, alpha- and beta-neo-endorphin, and a number of peptides derived from proenkephalin [14, 51–55]. The basic approach of a receptor-binding assay involves a drug molecule tagged with a reporter group (usually a radioactive atom) and detects the receptor in a tissue extract by rapidly separating the receptor-bound drug from the free unbound drug, usually using a filtration method to isolate the membrane-bound receptor [56]. If an endogenous substance is present in a tissue extract, the addition of this extract causes a decrease in drug binding due to competition between the endogenous molecule and the labeled drug for the receptor (Figure 8). The tissue extract is then fractionated using standard chromatography techniques and each fraction tested again in the receptor-binding assay. This process is repeated until the endogenous substance is pure, and then if it is a peptide, it is identified by sequencing (originally by Edman degradation and more recently by mass spectrometry).

This approach led to the discovery of many neuropeptides, including the opioid receptor-binding peptides described above. However, the requirement for an appropriately labeled drug that can be used for a receptor-binding assay has limited the utility of this technique. When techniques were developed in the 1980s to manipulate and sequence DNA, it became clear that there were a large number of receptor-like proteins for which there was no known drug. Additional approaches were developed to search for ligands of these putative receptors, dubbed "orphan" receptors. In the

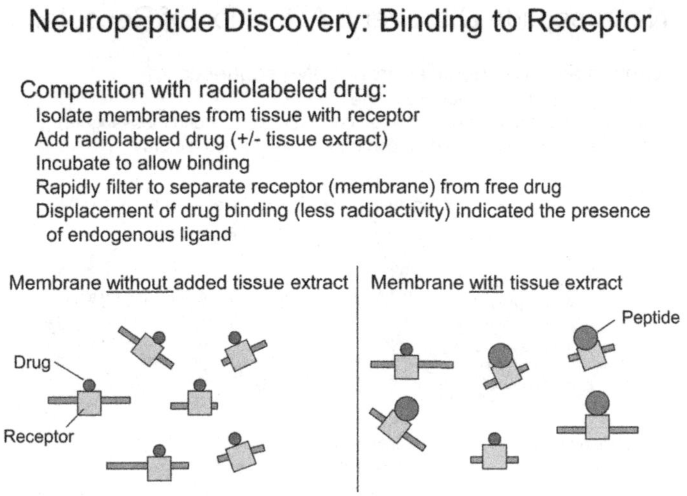

Neuropeptide Discovery: Binding to Receptor

Competition with radiolabeled drug:
Isolate membranes from tissue with receptor
Add radiolabeled drug (+/- tissue extract)
Incubate to allow binding
Rapidly filter to separate receptor (membrane) from free drug
Displacement of drug binding (less radioactivity) indicated the presence of endogenous ligand

Membrane without added tissue extract
Membrane with tissue extract
Peptide
Drug
Receptor

FIGURE 8: General principle of the radioligand receptor-binding assay. Left: The drug labeled with a radioactive tag (red circles) is allowed to bind to membrane homogenates containing receptor proteins (blue boxes). Right: Replicate tubes containing tissue extract are mixed with the same amount of radiolabeled drug and membrane homogenates. If the tissue extracts contain a peptide (purple circles) that binds to the same receptor site as the drug and the receptors are saturated with drug and peptide, the total amount of radioactive drug bound to receptor will be lower in the presence of the peptide. Following equilibration, the samples are filtered to separate receptor-bound drug from free drug (not shown) and then the amount of radioactivity bound to the filter is measured. Reduced radioactivity indicates the presence of a substance that competes for the drug-binding site.

1980s and 1990s, a large amount of effort was required to identify receptor DNA sequence. Now that the genomes of many species have been determined, it has become much easier to identify orphan receptor sequences based on homology searches. There are currently estimated to be dozens of orphan receptors that are likely to bind neuropeptides. However, many of these may instead bind small molecules; it is hard to accurately predict whether a G-protein-coupled receptor will be activated by peptides or small molecules (i.e., dopamine, serotonin, or related substances).

The basic approach to identify the endogenous ligand for orphan receptors involves the expression of the receptor in a cell line that also expresses common second messenger systems (Figure 9). Then, tissue extract is added and assays performed to look for activation of these second messengers. Cells exposed to a receptor agonist will show an increase in the signal. Because cells normally express some of their own receptors, it is important to control for this by also including the wild-type cells in the assay with tissue extract. A greater response by the cells expressing the specific receptor, compared with non-expressing cells, is taken as a reflection that the tissue extract contains

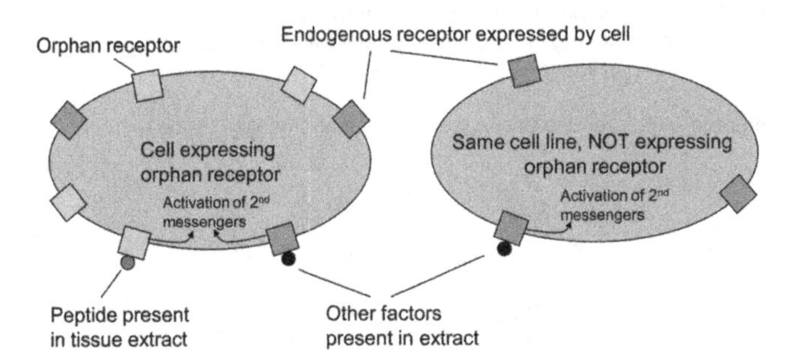

FIGURE 9: Neuropeptide discovery based on binding to a receptor expressed in a cell line. Orphan receptors identified by bioinformatics screens based on DNA sequences are expressed in cells along with second messengers that are likely to couple to these receptors. Control cell lines in which the orphan receptors (blue squares) are not expressed serve as the control for signaling through the receptors that are endogenous to the cells (orange squares). The cells are exposed to tissue extracts and screened for activation of the second messenger pathways. If the receptor-expressing cells (left) respond differently than the control cells (right), it means that a factor is present which binds to the expressed receptor.

an endogenous ligand for that receptor. As with the other approaches described above, once an initial signal is observed, the extract is fractionated, each fraction retested, and the process repeated until the compound is pure and can be identified. And, as with these other approaches, there is no guarantee that this process will result in the identification of neuropeptides; this approach has led to the discovery of many small molecules as well as many neuropeptides, including nociceptin/orphanin [57, 58] and orexins A and B [59].

The approach using receptor expression will only detect agonists, and will not detect peptides that are receptor antagonists unless an agonist has already been identified and is used in the assay. However, the vast majority of known neuropeptides are receptor agonists and only a few are antagonists: examples include agouti gene-related peptide, one form of beta-endorphin, and a short form of hemopressin [60–63]. On the other hand, it is possible that many more neuropeptide antagonists exist, and the small number of known examples is simply a reflection of the bias of the screening assays to detect receptor agonists.

2.3 NEUROPEPTIDE DISCOVERY USING GENETIC APPROACHES

In some model organisms, neuropeptides have been discovered by genetic approaches in which gene mutations were created, the organisms screened for an unusual phenotype, and the gene subsequently identified [64]. Because this approach will detect mutations in many different genes, it is not an approach to use if the goal of the study is the identification of neuropeptides. The genetic approach has also been used in several mammalian species to detect mutations in genes encoding neuropeptides or their receptors, and although this did not lead to the discovery of the neuropeptide, in some cases, the genetic mutation led to the discovery of additional functions for the neuropeptide. For example, positional cloning of a gene that caused narcolepsy in dogs led to the finding of the gene encoding hypocretin/orexin receptors [65]; the hypocretin/orexin peptides and their receptors had already been identified but thought to play a role in feeding [59, 66]. The narcoleptic dogs clearly showed that the orexin system was involved in arousal, which was subsequently confirmed in mouse models and also found to occur in humans [67–70]. In the case of humans, narcolepsy commonly results from loss of hypocretin /orexin-expressing cells rather than a mutation in hypocretin/orexin neuropeptide or receptor genes [68–71].

An example of a cell-cell signaling molecule found by genetic approaches in mice is leptin, which was discovered by positional cloning of the *obese* gene [72]. As the name implies, the *obese* mutation caused the mice to be extremely overweight. Leptin is a small protein secreted from adipose cells via a constitutive pathway. Plasma levels of leptin reflect the amount of body fat in mammals. Although leptin is a small endocrine protein, too large to be considered a peptide, it does play a critical role in controlling body weight through activation of specific hypothalamic neurons that express leptin receptors and the downstream mediators of leptin signaling include a number of neuropeptides [73–75].

While genetic approaches provide a great deal of functional information, they are limited by many factors. First, as mentioned above, they do not specifically detect neuropeptides but will find any gene involved in the pathway of production, secretion, or signaling and only a very small fraction of these are neuropeptides. Second, genetic analyses can take years. Finally, a large element of luck is required when looking for naturally occurring mutations.

2.4 NEUROPEPTIDE DISCOVERY USING PHYSICAL PROPERTIES: PRESENCE OF C-TERMINAL AMIDE

The methods described above detect any molecule involved in cell-cell signaling and are not restricted to neuropeptides. The first method to specifically look for neuropeptides was based on a property shared by many neuroendocrine peptides; the presence of a C-terminal amide. Proteins normally end with a carboxyl group, and peptides that represent proteolytic breakdown fragments of proteins also contain free C-terminal carboxyl groups. In contrast, many peptide hormones and some neuropeptides have a modified C-terminus, which consists of an amide group (Figure 10). The C-terminal amide group is synthesized by two enzymatic activities, which are collectively named peptidyl alpha-amidating monooxygenase (PAM); this amidation step is described in more detail in Chapter 3. In the early 1980s, Tatemoto and Mutt [76] reasoned that because the presence of a C-terminal amide was unique to neuroendocrine peptides, a search for novel amidated peptides would reveal unknown cell-cell signaling peptides. They used a chemical method to screen extracts of pig intestine for the presence of an amidated amino acid. The extract was fractionated, and each fraction was tested for the presence of an amide group. Those fractions containing C-terminally amidated amino acids were further fractionated, and the process was repeated until the peptides were of sufficient purity for identification by amino acid sequencing. Mutt, Tatemoto, and coworkers used this approach to identify neuropeptide Y and several other neuroendocrine peptides; some of these amidated peptides are neuropeptides that function in the brain (such as neuropeptide Y) while others play a hormonal role [77–80].

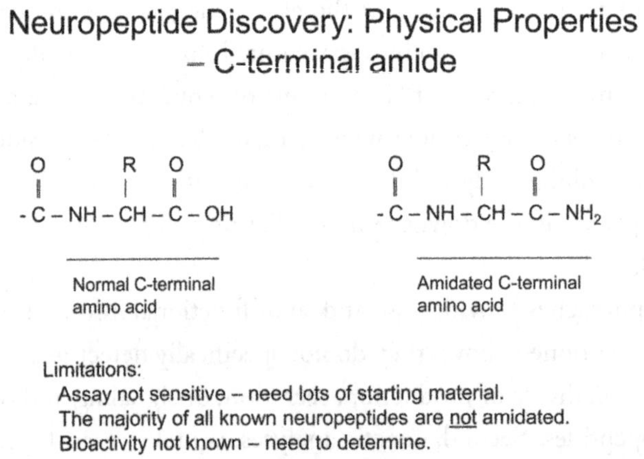

Neuropeptide Discovery: Physical Properties
– C-terminal amide

O R O
-C – NH – CH – C – OH

Normal C-terminal amino acid

O R O
-C – NH – CH – C – NH_2

Amidated C-terminal amino acid

Limitations:
Assay not sensitive – need lots of starting material.
The majority of all known neuropeptides are not amidated.
Bioactivity not known – need to determine.

FIGURE 10: Comparison of C-terminal carboxyl groups and C-terminal amide groups. The C-terminus of a normal protein or peptide contains a carboxylic acid, which readily ionizes into the negatively charged form at physiological pH. In contrast, the C-terminal amide group is uncharged at neutral pH. To detect peptides that contained amino acids with a C-terminal amidated residue, the sample was enzymatically digested, the resulting amino acids were labeled with a fluorescent tag, and analyzed using thin-layer chromatography that was able to separate amidated and non-amidated amino acids [76]. Samples containing an amide were further fractionated and the assay was repeated on the resulting fractions until a single peptide was purified and its sequence determined.

Although this approach is a fairly direct way to detect neuroendocrine cell-cell signaling peptides, there are many limitations. First, the assay used by Mutt and Tatemoto was not sensitive. Their initial discovery of neuropeptide Y required ~5000 kg of pig intestine as a starting material [76]. The scale of the extraction of this amount of tissue is beyond the capacity of most laboratories and requires a Herculean effort. Second, the majority of brain neuropeptides are not C-terminally amidated (see Chapter 5). Although this modification is fairly common in peptide hormones that require stability in plasma, it is less critical for neuropeptides that are secreted from neurons to signal neighboring neurons. A third problem is one that is common to all of the physical approaches; the functions of the peptides identified by these approaches are not known and required many additional studies. In the case of neuropeptide Y, it was years after the initial discovery before the peptide was tested in a bioassay and found to dramatically elevate food intake [81, 82].

2.5 NEUROPEPTIDE DISCOVERY BASED ON THE SPECIFIC PROPERTIES OF THE PEPTIDE'S PRECURSOR

With the development of techniques to manipulate DNA in the 1980s, it became possible to identify potential neuropeptide precursors from analysis of the properties of the protein encoded by the DNA. These properties include the requirement of an N-terminal signal peptide domain that directs the protein into the endoplasmic reticulum. A typical neuropeptide precursor is small, usually 300 amino acids or fewer, although some larger proteins appear to represent neuropeptide precursors (i.e., proteins named VGF, chromogranins A and B, and secretogranin II). Neuropeptide precursors also contain specific cleavage sites for processing by endopeptidases and exopeptidases. These processing sites are typically pairs of basic amino acids, commonly Lys-Arg and Arg-Arg (see Chapter 3). Finally, a comparison of the sequence among species usually reveals highly conserved domains that often represent the bioactive peptides.

One of the best examples of a neuropeptide discovered by this approach is calcitonin gene-related peptide (CGRP). Calcitonin is a peptide hormone produced in the thyroid. Soon after the identification of the calcitonin gene, it was found that a related mRNA was expressed in the brain, with high levels of expression in the hypothalamus [83, 84]. This mRNA was found to represent an alternatively spliced form of the calcitonin gene in which exon 4 was missing, and instead two additional exons were present (Figure 11). Protein produced from the brain transcript shared the N-terminal half of the calcitonin precursor but differed immediately preceding the cleavage site used to generate calcitonin and instead encoded a distinct cleavage site, a peptide of 37 amino acids, another cleavage site, and a short C-terminal peptide of 4 amino acids. Further studies found that the 37-residue peptide, CGRP, functions as a neuropeptide in the brain and the periphery [85].

When it was discovered that the calcitonin gene was differentially spliced to give rise to a distinct neuropeptide, it was predicted that this would be a common mechanism for the generation

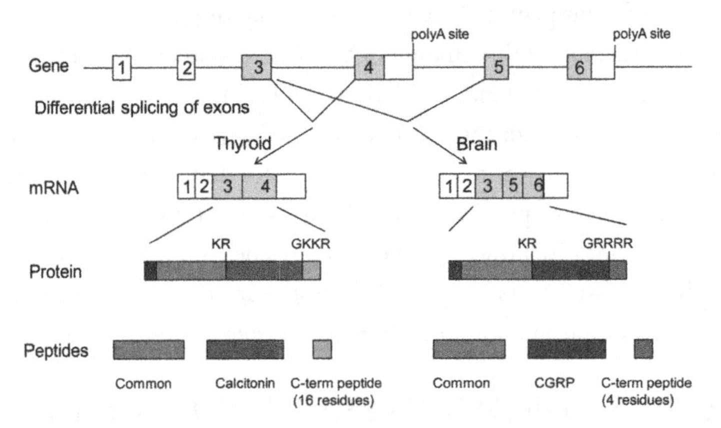

FIGURE 11: Calcitonin and calcitonin gene-related peptide (CGRP). A single gene encodes both calcitonin and CGRP. Common exons include the non-coding exons 1 and 2 and exon 3, which encodes the AUG translation initiation site, signal peptide, and N-terminal 51 residues of the prohormone. Exon 4 contains the Lys-Arg cleavage site (KR), the calcitonin sequence, the second cleavage site (GKKR), and then a short 16-residue C-terminal peptide. This C-terminal peptide was found to exist together with calcitonin in thyroid cells but a function for this peptide has not been conclusively identified [86]. In the brain and peripheral neurons, the gene is spliced such that exon 4 is deleted and a downstream 3′ splice acceptor site on exon 5 is used, which is spliced to exon 6 [87]. The resulting protein shares the N-terminal half with calcitonin but has a unique processing site (also KR, but with different surrounding residues) followed by the 37-residue peptide sequence of CGRP, another processing site (GRRRR), and a short 4-amino acid-long C-terminal peptide.

of multiple bioactive peptides [84]. However, in the 3 decades since this discovery, the vast majority of neuropeptide and peptide hormone-encoding genes have not been found to undergo differential splicing, and this remains a rare mechanism for the production of distinct peptides. On the other hand, many peptides have been identified based on precursor analyses and predictions of cleavage sites. In the past decade, bioinformatics approaches have been used to scour databases of genomic and cDNA sequences to search for potential neuropeptide precursors. Previously, precursors were found during searches for genes with unique properties. For example, the neuropeptide precursor named cocaine- and amphetamine-regulated transcript (abbreviated as CART) was identified during a search for mRNAs up-regulated by treatment of rats with either cocaine or amphetamine [88]. This gene is one of the small number of neuropeptide genes that show differential splicing, and produces translation products of either 116 or 129 amino acids [88]. Peptides derived from both forms of CART mRNA have been identified, and the peptides are thought to function as a neuropeptides involved in feeding and body weight regulation [89]. Another example of this approach to peptide discovery is the peptide named hypocretin; this peptide was also discovered using the orphan receptor approach described above and named orexin. A cDNA encoding the hypocretin precursor was found in a search for mRNAs specifically expressed in the hypothalamus and not in other brain regions [66]. In both cases, the fact that the mRNA encoded a neuropeptide precursor was a matter of chance, as there are many other genes up-regulated by cocaine and amphetamine or which show regional differences in expression in rodent brain. Once the potential neuropeptide precursor was identified, it was necessary to show that the protein was expressed in neurons and processed into peptides within the regulated secretory pathway and then secreted from the cells upon stimulation. As with other peptide-first approaches, it was also necessary to determine which, if any, of the peptides produced from the precursor had biological activities and then determine their function.

A related approach to discover novel neuropeptides makes use of the principle that most bioactive peptides are made from precursors that encode multiple bioactive moieties. In this approach, the precursors of known neuropeptides are examined for potential cleavage sites, based on the established specificities of the precursor-cleaving enzymes (described in detail in Chapter 3). A comparison of the sequence among species can also reveal potential neuropeptides; these regions are often highly conserved among species, whereas the non-functional portions of the precursor usually show lower levels of conservation. Peptides corresponding to these domains are synthesized and tested for biological activity in assays, such as receptor-binding techniques, bioassays with tissues, and/or studies with live animals. It is also necessary to show that the peptide is actually produced in a biological system; not all predicted cleavage sites are used, and processing can occur at sites that are not predicted (see Chapter 3).

A major difficulty in this approach, which is common to all "peptide-first" methods, is deciding which of the many potential assays should be used to test the peptide identified. Knowledge of the distribution of the peptide can help guide the choice of bioassay to be tested. In addition, when looking for novel bioactive peptides within the precursors of known neuropeptides, it is common to test the novel peptides with the same assays as the known peptide; in some, although not in all cases, the various peptides produced from a single precursor have related biological properties.

2.6 NEUROPEPTIDE DISCOVERY USING PRECURSOR-PROCESSING ENZYMES

Most classical neuropeptides are produced from precursors that require processing by endopeptidases followed by removal of the basic C-terminal residues by a carboxypeptidase (Figure 12). Because of the widespread involvement of CPE in peptide processing, a strategy to isolate substrates of this enzyme was developed to identify novel neuropeptides [90]. The first part of this scheme

Identification of neuropeptide processing enzyme substrates

- Step 1: Block carboxypeptidase E – accumulate peptide-processing intermediates

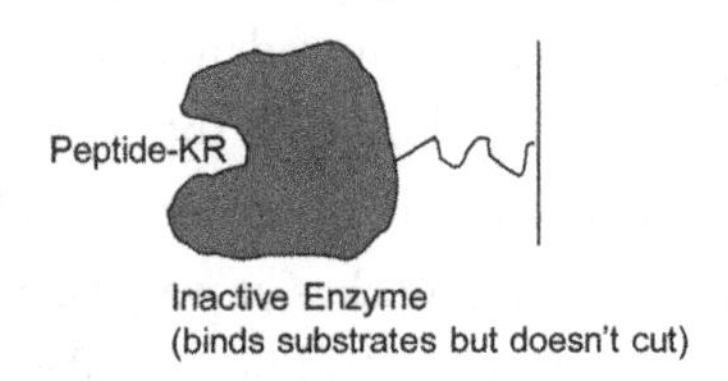

- Step 2: Purify peptide-processing intermediates on affinity column

Peptide-KR

Inactive Enzyme
(binds substrates but doesn't cut)

- Step 3: Elute peptide-processing intermediates from column and identify by mass spectrometry "peptidomics" approach

FIGURE 12: Strategy to identify neuropeptides based on the involvement of CPE during peptide biosynthesis. Step 1 requires the inhibition of CPE. Mice with an inactivating mutation in the *Cpe* gene accumulate CPE substrates. Step 2 involves the purification of peptides on an affinity resin, which binds peptides with C-terminal basic residues. Anhydrotrypsin-agarose binds peptides only with C-terminal basic residues and not internal basic residues (unlike trypsin, which preferentially binds peptides with internal basic residues). Step 3 involves the elution of the peptides from the affinity column and identification using mass spectrometry.

required the inhibition of CPE; normally this enzyme is not the rate-limiting enzyme in neuropeptide production and the CPE substrates are present in the brain at extremely low levels, relative to the levels of the mature neuropeptides. Therefore, to detect CPE substrates it was necessary to inhibit the enzyme so that the intermediates with C-terminal basic residues could accumulate. Initial attempts to accomplish this using potent inhibitors of CPE were unsuccessful, presumably because these inhibitors were highly charged and did not gain access to CPE within the secretory vesicles; this approach requires the inhibitor to first enter the cell and then cross into the secretory vesicle. Fortunately, this approach became feasible upon the discovery that the *fat* mouse contains a point mutation within the coding region of CPE that results in inactivation of the enzyme and accumulation of neuropeptide processing intermediates with C-terminal basic amino acids [91]. Rather than inhibit CPE with chemical compounds, it was possible to simply use *fat* mouse brain as the starting material.

The second step in this scheme was to purify the peptide processing intermediates on an affinity column consisting of an inactive peptidase. Enzymes have two properties: first, they bind substrates, then they catalyze the conversion of substrate into product(s). The initial idea was to use a point mutation of CPE in which the critical active site Glu was converted into a Gln residue, which was predicted to be inactive (this is not the same mutation as the one found in *fat* mice). Although this approach worked in that CPE with the Glu270Gln mutation was completely inactive as a peptidase but was still able to bind to peptides with C-terminal basic residues, it was difficult to produce sufficient amounts of the mutant enzyme and couple it to a resin without destroying the peptide-binding properties. Fortunately, an affinity matrix that bound peptides with C-terminal basic residues was commercially available, consisting of agarose resin attached to trypsin in which the active site Ser was chemically converted into the anhydro form (i.e., lacking the critical -OH group of Ser). Trypsin normally has higher affinity for peptides containing internal Lys or Arg residues (which are substrates) than for peptides with these groups on the C-terminus (which are products). However, testing of the anhydrotrypsin resin showed that the conversion of trypsin to the anhydro form eliminated substrate binding but not product binding, and the resin bound only peptides with C-terminal Lys or Arg residues and not peptides with internal Lys or Arg residues. This allowed the anhydrotrypsin column to be used for the purification of CPE substrates that accumulated in the *fat* mouse.

The final step in this scheme was to sequence the affinity-purified peptides from the *fat* mouse tissues, comparing the results with those from similarly processed wild-type mouse tissues. This step used mass spectroscopy approaches similar to those developed for proteomics, except that unlike proteomics techniques that digest the protein with trypsin to form peptides, peptidomic approaches skip this enzyme digestion so that information on the native form of the peptide can be obtained. This resulted in longer peptides than typically found from tryptic digests of proteins, and the longer peptides were often difficult to sequence using tandem mass spectrometry. However, improvements in mass spectrometry instruments and in the computer programs for database searching led to the identification of a large number of peptides [90, 92].

Overall, hundreds of peptides were identified by this technique. Most of these corresponded to either known neuropeptides, which validated the technique, or to novel fragments of known neuropeptide precursors; the latter group may represent neuropeptides. However, many of these neuropeptides were predicted from bioinformatic analysis of the precursor and the known specificity of the peptide-processing enzymes, and so their identification merely confirmed the cleavage site predictions and showed that the peptides actually existed in the brain. In addition to these fragments of known neuropeptides, several peptides were identified that did not match any sequence present in the database of known proteins [92]. Two of these corresponded to the mouse homolog of peptides previously identified from sequence analysis of peptides isolated from bovine adrenal chromaffin granules by Humberto Viveros and colleagues [93]. The novel mouse peptides were named SAAS, KEP, GAV, LEN, and PEN, based on amino acid sequences present within each of the peptides [92]. In addition, two forms of SAAS were detected: big SAAS and little SAAS. Subsequent analysis revealed many additional forms of each of these peptides, as is typical for neuropeptides [32]. Further studies aimed at establishing the function of the proSAAS peptides are described in Chapter 5.

The above approach to identify neuropeptides based on the accumulation of peptide-processing intermediates in the CPE-mutant *fat* mice has several advantages over other methods. First, the method is selective for neuropeptide precursors. Although some of the non-neuropeptides present in tissue extracts contain C-terminal Lys or Arg, these peptides are present in both the *fat* and wild-type mouse tissue extract, whereas the CPE substrates are present at much higher levels in the *fat* mouse extracts. By comparing mass spectrometry analyses of peptides present in wild-type mouse brain extracts with comparable extracts from *fat* mice, it was possible to find those peptides greatly enriched in the *fat* mice and focus on them, rather than the peptides common to both samples. Second, because the approach uses mass spectrometry, it is extremely sensitive and able to detect peptides extracted from a single mouse pituitary; this is over a billion times less tissue than used for the chemical assay to detect C-terminally amidated peptides by Tatemoto and Mutt [76]. (Although to be fair, if the chemical method to detect C-terminally amidated peptides were to be attempted today using modern mass spectrometry methods to sequence the peptides, it would require much less tissue than used by Tatemoto and Mutt in the early 1980s.) In addition to these advantages, there are also several limitations to the CPE-based approach. First, not all neuropeptides require CPE for their production; for example, the C-terminal peptide of each neuropeptide precursor usually does not require CPE (unless the precursor ends in C-terminal basic residues, which is rare). However, this is not a serious limitation because the technique will likely detect other pieces of the precursor and the C-terminal peptide can be predicted from the sequence of the precursor. For example, the proSAAS-derived peptide named big LEN, which represents the C-terminus of proSAAS, was not found in the initial analysis using anhydrotrypsin columns and was only identified in subsequent studies that examined peptides isolated without this step. A more serious problem is that mass spectrometry does not detect all peptides. Some peptides do not ionize well and show weak signals in the mass spectrometer. Furthermore, not all peptides can be sequenced by tandem mass spectrometry. For example, a large number of peptides have been detected by the affinity column/mass spectrometry approach but not yet identified by sequencing [32]. These unidentified peptides either represent completely novel peptides or known peptides with posttranslational modifications. The computer database searches start by calculating the mass of the observed peptide and then consider as candidates only those peptides that match the observed mass (within the error of the instrument used, which is typically less than 0.1 Da). If the peptide has a posttranslational modification that was not considered in the database search, the identity of the peptide cannot be established from this analysis. Another problem with mass spectrometry is that it can be difficult to detect peptides present in low abundance due to the background from the more abundant peptides. Therefore, peptides present in low levels or in a relatively small population of neurons (such as the hypocretin/orexin peptides) will be difficult to detect using mass spectrometry. Lastly, this approach suffers from the same problem as the other peptide-first approaches; additional studies are required to learn the function of the novel peptides.

2.7 NEUROPEPTIDE DISCOVERY USING PEPTIDOMIC APPROACHES

The peptide-first approaches described above will detect only classical neuropeptides; those produced in the secretory pathway and cleaved by conventional neuropeptide-processing enzymes. To detect non-classical neuropeptides, as well as all other peptides present in a tissue extract, mass spectrometry-based peptidomic approaches have been used. These approaches are related to the approach described above to detect substrates of CPE, except that there is no affinity column step; the tissue extract is directly applied to the mass spectrometer (Figure 13). Thus, both classical and

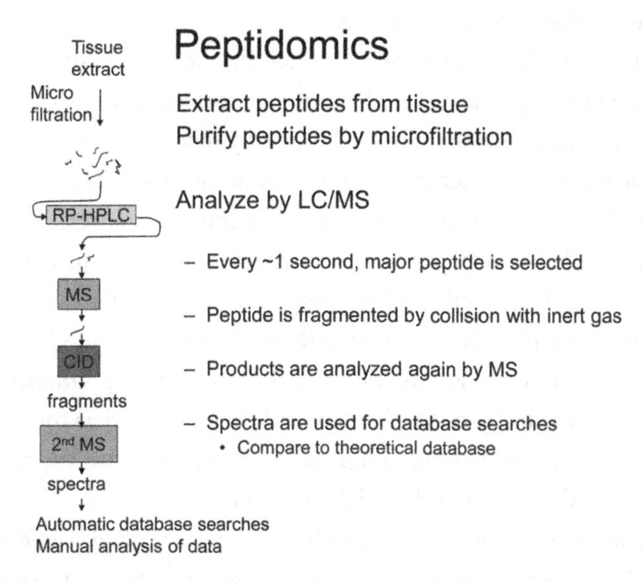

FIGURE 13: Peptidomics: the analysis of peptides in a biological sample. The general approach to peptidomics is similar to proteomics approaches that analyze peptides using liquid chromatography and mass spectrometry (LC/MS). However, for proteomics, the peptides are usually generated by digestion of proteins with trypsin or another protease, whereas for peptidomics, there is no enzyme digestion; this allows peptidomics studies to detect the endogenous forms of peptides present in tissues. Abbreviations: RP-HPLC, reverse-phase high performance liquid chromatography; MS, mass spectrometry; CID, collision-induced dissociation.

non-classical neuropeptides can be detected. This approach has led to the discovery of several forms of hemopressin, which are putative non-classical neuropeptides [20, 63, 94]. The hemopressins are described in more detail in Chapter 5.

Strengths of the peptidomic approaches are the relatively unbiased detection of the major peptides present in a tissue extract, great sensitivity, and speed of the process. Hundreds of peptides can be detected in a matter of hours—although determining the sequence of these peptides (and confirming the predictions from the database-searching programs) can take days or weeks of careful analysis. Although not biased by the requirement for CPE activity, as in the CPE substrate approach described above, the general peptidomics approach is still limited by the problem that not all peptides ionize well in the mass spectrometer. Furthermore, to be detected by mass spectrometry, the peptides must have at least one positive charge on the peptide when run in positive ion mode, which is the usual way that mass spectrometry is performed. Some peptides contain a blocked N-terminus (such as peptides modified by acetylation or pyroglutamylation), and unless there is an internal Lys or Arg, these peptides will not be detected. Finally, the common problem of the peptide-first approaches applies to this peptidomics approach, and additional testing is necessary to determine the function of the identified peptides.

A recent adaptation of this peptidomics method has been to study those peptides secreted from a tissue or cell line [18, 95]. The idea is that only the secreted peptides are likely to function as either classical or non-classical neuropeptides. This greatly reduces the number of peptides requiring further testing for biological function. Another variation of the basic peptidomics technique is to perform quantitative analysis of the relative level of peptides between samples. This provides information on the regulation of a peptide, or relative levels in different brain regions; these can be helpful in understanding the function of the peptides. One quantitative approach involves the postextraction labeling of the peptides using trimethylaminobutyric acid (TMAB) containing different numbers of deuterium or ^{13}C atoms (Figure 14). After labeling, samples are pooled and the peptides separated from proteins by microfiltration and then analyzed as for standard peptidomics approaches. The resulting data show peak groups for each peptide, with the relative peak intensity reflecting the level of peptide in each of the original samples.

Other approaches to peptide quantitation have been developed, some using isotopic tags. In addition to the TMAB tags, reagents used for labeling peptides that can be detected in MS include acetic anhydride, succinic anhydride, formaldehyde, and other reagents [96–99]. Another related reagent is iTRAQ, which also labels amines but which can only be quantified from MS/MS spectra, not from the primary MS data [100]. Each of these approaches has both advantages and disadvantages. The commercial availability and relatively low cost of acetic and succinic anhydride is offset by the lack of co-elution of the isotopic forms from the high-resolution HPLC columns. Furthermore, these labels convert a positive charge into a negative (succinate) or neutral residue

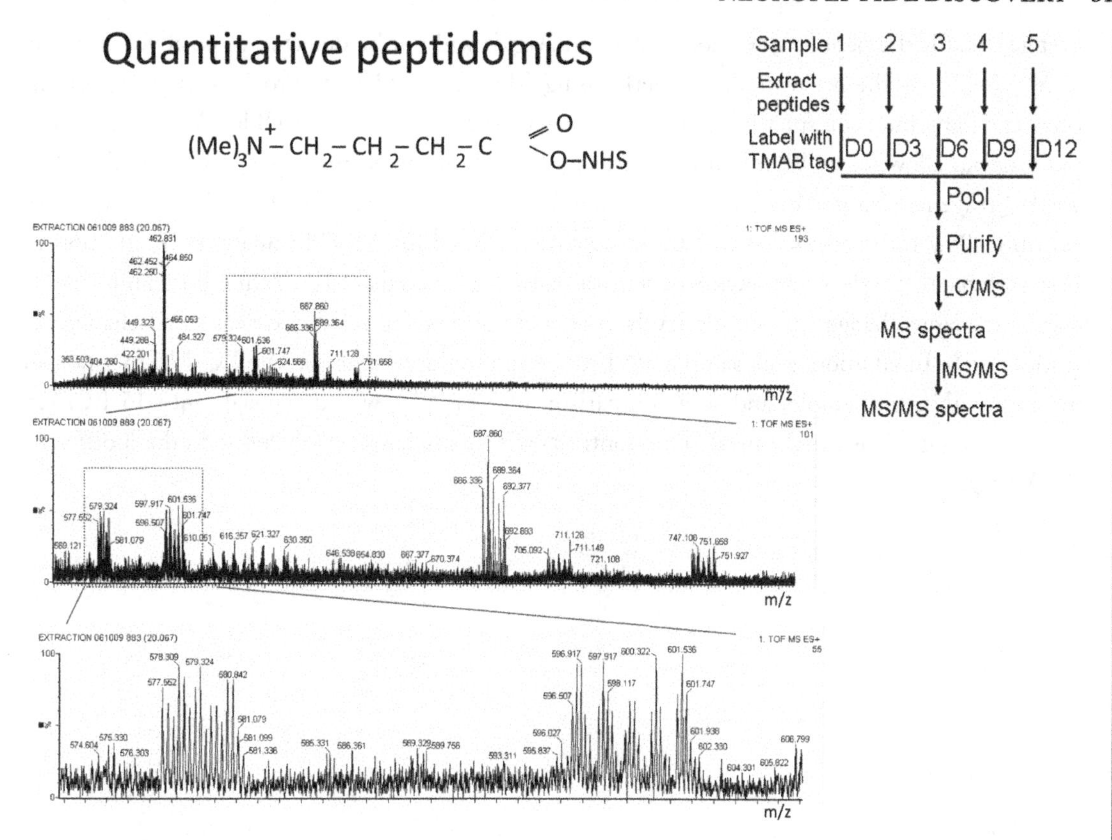

FIGURE 14: General quantitative peptidomics approach. A variation of the general peptidomics approach is to label the extracted peptides with one of several similar reagents that react with free amine groups on the peptide (N-termini and Lys side chains). One such set of reagents is TMAB-N-hydroxysuccinimide (TMAB-NHS). This reagent can be synthesized with methyl groups containing ^{12}C and hydrogen (D0), or with methyl groups containing from 1 to 3 atoms of deuterium; because 3 methyl groups are incorporated into the compound, the resulting masses are 3, 6, or 9 Da higher than the D0-TMAB reagent. In addition, a compound with methyl groups containing 3 deuteriums and ^{13}C can be generated to produce a reagent that is 12 Da heavier than the D0 reagent. After labeling with the various reagents, the samples are pooled, the peptides are purified away from salts and proteins, and then the peptides are analyzed on LC/MS. A representative spectrum is shown, which corresponds to a single 1-second-long accumulation of ions. Spectra taken several seconds later or earlier show completely different patterns of peptides. The top spectrum represents the full mass range analyzed (mass/charge, or m/z ratio of 300–1800), and the lower spectra represent computer-generated highlights of specific mass ranges. Note the presence of 5 distinct peaks for each of the samples, with the peak intensity corresponding to the levels of peptide present in the original sample.

(acetate), and if the peptide does not contain an Arg, it will not be detected in positive ion mode on MS (which is the usual mode for performing MS). Formaldehyde provides only two forms, heavy and light, but is inexpensive and relatively easy to perform. The TMAB labels require custom synthesis but allow for multiple isotopic forms and maintain the positive charge of the amine. The iTRAQ reagents also provide multiple isotopic forms for multivariate analyses, but are expensive and only allow for quantification of those peptides selected for MS/MS analysis. In addition to these isotopic methods, some scientists perform label-free analysis [101]. While this can be useful to examine large changes in peptide levels, it is more difficult to reliably detect small changes in peptide levels. In addition, each sample needs to be run multiple times on the LC/MS to establish variation within the sample, and so a comparison of 5 samples would translate into 15 LC/MS runs (if each sample is run 3 times). This contrasts with a single run if labeled with the 5 different TMAB tags.

• • • •

CHAPTER 3

Neuropeptide Biosynthesis

The biosynthetic pathways for classical neuropeptides were largely established in the 1980s and 1990s, although a few gaps remain, as discussed below. In contrast, little is known about the biosynthetic pathways for non-classical neuropeptides.

3.1 BIOSYNTHESIS OF CLASSICAL NEUROPEPTIDES: EARLY EVENTS IN THE ENDOPLASMIC RETICULUM AND GOLGI

Classical neuropeptides are produced from precursor proteins by a series of post-translational modifications that are catalyzed by a number of proteolytic and non-proteolytic enzymes. The first enzymatic step in the biosynthesis of classical neuropeptides is the peptidase that removes the N-terminal domain that directs the nascent protein into the lumen of the endoplasmic reticulum [102, 103]. This enzyme, named the signal peptidase, is not specific for neuropeptide precursors; it removes the N-terminal signal peptides from all proteins that contain this domain such as peptide processing enzymes, lysosomal enzymes, and many membrane-bound proteins. After the removal of the N-terminal signal peptide, the peptide precursor is folded in the endoplasmic reticulum through interactions with one or more of several chaperone proteins [104]. If the precursor contains an appropriate sequence for the addition of N-linked glycosylation (Asn-Xaa-Ser/Thr, where Xaa is any amino acid except Pro), complex carbohydrate chains are attached to the Asn residue in the endoplasmic reticulum [105]. These carbohydrate chains contain a total of 14 sugar molecules, consisting of 2 N-acetylglucosamines, 9 mannoses, and 3 glucoses. Following the addition of the core sugar block by an oligosaccharyltransferase, the glucose residues are removed by a glycosidase and the protein is exported to the Golgi, where further processing occurs. This processing includes both removal and addition of carbohydrate units to create complex side chains that are often tissue-specific. O-linked carbohydrates may also be attached to Ser or Thr residues in the Golgi; the consensus sequence for their addition is not known. Sulfation of Tyr residues also occurs in the Golgi [106]. This addition, which occurs in the vicinity of acidic groups, is found on a small number of neuropeptides such as cholecystokinin (CCK), where three Tyr undergo this modification. One of these sulfation events is within the bioactive CCK moiety, and the presence of a sulfo-Tyr affects the biological activity of the peptide [107]. Phosphorylation of Ser or Thr residues is another uncommon modification of neuropeptides; examples include adrenocorticotropic hormone (ACTH) and the ACTH-derived peptide named corticotropin-like intermediate lobe peptide (CLIP) and a peptide derived from the C-terminal region of proenkephalin [108]. Unlike sulfation, the presence of the phosphate group is not known to influence the biological activities of these peptides. The enzyme responsible for the phosphorylation of neuropeptides has recently been identified [296]. This activity has been termed casein kinase 2-like based on its recognition sequence of Ser/Thr-Xaa-Asp/Glu. As a side point, the requirement for the Asp or Glu in the downstream position is not absolute and can be substituted by a phosphoSer/Thr residue, as occurs in the phosphorylation of the C-terminal region of proenkephalin [109].

The movement of the neuropeptide precursor from the endoplasmic reticulum to the Golgi and then movement through the Golgi occurs via small vesicles that shuttle the contents of the organelles from one compartment to another [110]. This process requires cytosolic and membrane-

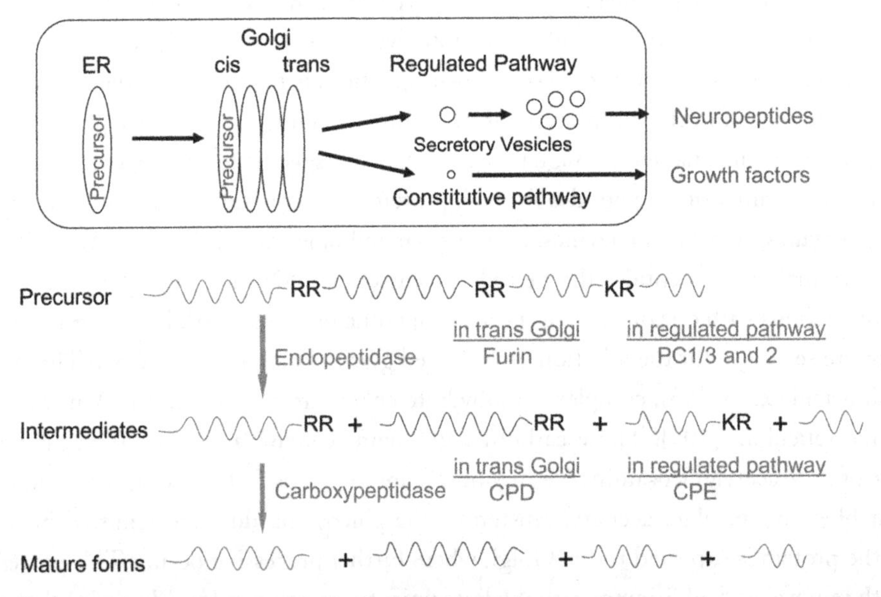

FIGURE 15: Classical neuropeptides are produced by selective cleavage of precursors within the secretory pathway. For certain neuropeptides, proteolytic processing begins in the trans-Golgi network, located at the trans side of the Golgi. Peptidases in the trans-Golgi network include the endopeptidase furin and the exopeptidase carboxypeptidase D (CPD). For most neuropeptides, the processing within the regulated secretory vesicles, or granules, plays the major role in peptide biosynthesis. The secretory vesicle peptidases include prohormone convertase (PC) 1/3 and 2, and carboxypeptidase E (CPE).

bound factors including GTPases, guanine nucleotide exchange factors, and other proteins that form the specific coat protein complex that initiates the budding process (named COPII proteins). Some of the transmembrane proteins interact with the cargo and ensure the proper sorting of the luminal proteins to the correct intracellular location.

Many of the proteins involved in the post-translational modifications and trafficking mechanisms described above are common to a number of secretory pathway proteins as well as proteins that are routed to lysosomes or the plasma membrane. In the late Golgi, the neuropeptide precursor is routed into the trans-Golgi network where it is sorted into immature secretory granules. Some lysosomal enzymes also are sorted into these granules, but these are generally thought to be subsequently routed to lysosomes and not retained within mature secretory granules. Once sorted into the immature vesicles in the trans-Golgi network, maturation of these vesicles to exclude lysosomal and Golgi proteins occurs. The neuropeptide precursors may then be acted on by enzymes that are unique to the regulated secretory pathway (Figure 15). In the following sections, the general trans-Golgi enzymes are discussed first, followed by the neuropeptide-specific enzymes that function in the maturing secretory granules.

3.2 FURIN AND OTHER PROTEASES IN THE TRANS-GOLGI NETWORK

Once in the trans-Golgi network, the neuropeptide precursor encounters proteolytic enzymes that begin to cleave the precursor if the specific sites recognized by these proteases are present within the neuropeptide precursor (Figure 15). The major endoprotease of the trans-Golgi network is furin, which is a membrane-bound protease that cleaves peptides with the consensus site Arg-Xaa-Xaa-Arg, with cleavage occurring after the last Arg [111, 112]. A basic residue in the P2 position (i.e., Arg-Xaa-Lys/Arg-Arg) is preferred by furin. Other trans-Golgi network endopeptidases include certain isoforms of PACE4 and proprotein convertase 5. All of these enzymes have generally similar substrate specificities, cleaving on the C-terminal side of an Arg residue within a sequence that includes one or more additional basic residues upstream of the cleavage site [113].

Furin was the first of these mammalian endoproteases to be identified. This gene was originally identified in a search for oncogenes and was named "FUR" to stand for FES upstream region [114]. When the full-length cDNA was sequenced in 1989, it was noted that the sequence showed substantial homology to a yeast protease involved in the cleavage of alpha-mating factor at paired basic residues (which was named Kex2; also called kexin) [115, 116]. Therefore, it was considered likely that the FUR gene encoded a protease, and when further studies confirmed this, it was renamed furin, the "-in" being a common suffix for proteases (i.e., trypsin, chymotrypsin). Although initially considered a candidate for the major endoprotease involved in neuropeptide biosynthesis, the location of furin in the trans-Golgi network and not in the maturing secretory granules suggested another primary function, the processing of proteins that transit either the regulated or constitutive secretory pathway. This includes a number of proteins that are constitutively secreted (such as growth factors) or which remain bound to the cell surface (such as the insulin receptor). The broad distribution of furin in neuroendocrine and non-neuroendocrine tissues supports a more generalized function than just neuropeptide production. However, it is likely that furin contributes to the production of certain neuropeptides. One precursor with a number of furin-like consensus sites is proSAAS, which appears to be cleaved earlier in the secretory pathway than typical neuropeptide precursors, consistent with a role for furin in certain cleavages [92].

The domain structure of furin is generally similar to the domain structure of the other related trans-Golgi network proteases (Figure 16). All of these contain an N-terminal signal peptide followed by a pro-region that needs to be removed before the enzyme becomes enzymatically active [113]. The enzymatic domain has amino acid sequence homology to the yeast enzyme Kex2 and distant homology to the bacterial enzyme subtilisin. All of these enzymes are serine proteases. The active site Ser residue is more reactive than a typical Ser residue because in the three-dimensional structure the hydrogen ion of the –OH group is pulled off by a His residue that in turn is spatially close to an Asp residue. This Ser-His-Asp cascade is also found in endoproteases such as trypsin and chymotrypsin. However, there is no amino acid sequence homology between the subtilisin/kexin/furin group and the trypsin/chymotrypsin group, suggesting convergent evolution of these two enzyme families. In addition to the catalytic domain, furin and all other members of the furin gene family contain another domain, named the P-domain. In furin and some other family members, a Cys-rich domain follows the P-domain. Finally, furin and the other membrane-bound endoproteases contain a transmembrane binding region and cytosolic tail. Motifs in the cytosolic tail play an important role in the retention of furin in the trans-Golgi network and the movement of this protein from the trans-Golgi network to the cell surface and back through the endocytic pathway, eventually returning to the trans-Golgi network [117–119].

Furin is maximally active at neutral or slightly acidic pH values, reflecting the internal environment within the trans-Golgi network. Although only a small fraction of furin is present on the cell surface at any given moment, it is enzymatically active on the cell surface and able to cleave bacterial and viral proteins, in some cases activating these proteins and allowing the bacterial or viral proteins to become toxic and/or gain entry into the furin-expressing cell [112]. Furin may also contribute to processing of proteins taken up by cells in the endocytic system. A large number of furin substrates have been identified, and these include growth factors and growth factor receptors such as the insulin-like growth factor receptor, serum proteins such as blood clotting factors, extracellular matrix proteins such as matrix metalloproteases, and numerous other proteins including bacterial and viral proteins.

The crystal structure of furin has been determined [120] and is generally similar to that of Kex2 [121], as expected from the amino acid homology of the two proteins. The substrate binding

Furin

- 1986 – Gene and partial cDNA isolated
  - Named "FUR" gene for FES Upstream Region
- 1989 – Full length cDNA sequenced, predicted to be endopeptidase
- 1990s – Enzyme properties examined

Location:

- Broad tissue distribution
- Present in *trans* Golgi network (also cycles to cell surface)

Activity

- Cleaves after Arg-Xaa-Arg-Arg—
  Arg-Xaa-Lys-Arg—
  Arg-Xaa-Xaa-Arg—
- Active at neutral to slighty acidic pH (= TGN environment)

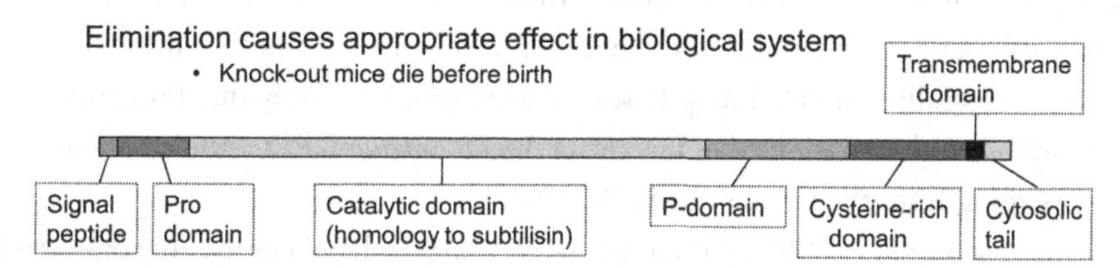

FIGURE 16: Key features of furin. The gene was originally named FUR for FES upstream region and renamed furin once its enzymatic properties were discovered. Furin is broadly expressed throughout the body, and within cells is most abundant in the trans-Golgi network. The optimal furin substrates contain basic residues in the P1, P2, and P4 positions, although the P2 position is less critical than the others (unlike the prohormone convertases). Mice lacking furin die during embryogenesis [122]. The domain structure of furin is indicated and discussed in the text.

region of furin contains pockets that bind to Arg in S1 and S4 positions of the substrate, and Lys or Arg in the S2 position, thus explaining furin's stringent substrate specificity.

3.3 CARBOXYPEPTIDASE D

Because furin and the related trans-Golgi network endoproteases cleave on the C-terminal side of the Arg residue, the furin product contains one or more C-terminal basic residues. These are subsequently removed by carboxypeptidase D (CPD), which like furin is located in the trans-Golgi network [123]. CPD is also attached to membranes via a transmembrane sequence near the C-terminus (Figure 17). Another similarity is that both furin and CPD have some common motifs in their C-terminal tails that project into the cytosol. The sequence similarity is too low to suggest divergent evolution from a common ancestor and instead is likely the result of convergent evolution. These common sequence motifs include acidic clusters with phosphorylatable Ser and/or Thr residues and di-Leu sequences; these motifs are involved in the retention of the proteins in the trans-Golgi network as well as their trafficking to the cell surface and reuptake via the endosomal pathway and return to the trans-Golgi network [124, 125]. Like furin, CPD is also broadly expressed throughout the neuroendocrine system and is also present in non-neuroendocrine cells [126, 127]. Although not present in maturing secretory vesicles, CPD still contributes to the biosynthesis of many neuropeptides, although for most neuropeptides this is only a minor role. These conclusions are based on analysis of the forms of peptides found in *fat* mice, which lack functional CPE activity [91]. In these mice, the levels of some neuropeptides are much lower than their levels in wild-type mice, while the levels of other peptides are not as severely affected by the absence of CPE activity [128]. For those peptides that require processing by a carboxypeptidase (i.e., all peptides other than the C-terminal peptide in the precursor), the presence of these peptides in *fat* mouse tissues indicates that another peptidase is responsible for their formation, and CPD is the most likely candidate. Unfortunately, this cannot be directly tested because mice with a disruption in the gene for CPD have not been successfully generated despite several attempts. It is likely that these mice would not be viable; *Drosophila* with mutations that inactivate the *Drosophila* CPD ortholog are not viable and die prior to emerging from the larval stage [129, 130].

CPD was discovered in 1995 by three groups working independently, only one of which was focused on carboxypeptidases; the other two were pursuing molecules identified from other studies. Kuroki, Ganem, and colleagues were interested in the liver cell surface proteins that bound to duck hepatitis B virus particles and brought the virus into the cells. They had previously identified a 180-kDa glycoprotein, named gp180, that was important for viral uptake [131]. Upon isolation of the cDNA encoding gp180 and sequence analysis, it became apparent that this protein was a member of the metallocarboxypeptidase family with similarity to CPE, except that gp180 contained three CP-like domains followed by a transmembrane domain and cytosolic tail [132]. A second group working on the *Drosophila* gene named *Silver* (*svr*) also found a multidomain carboxypeptidase [130]. The *svr* mutation had been discovered in the early 1900s and through positional cloning Settle et al. [130] mapped the location and sequenced the cDNA corresponding to the mRNA transcribed from the gene. However, due to a sequencing error, they only found two full carboxypeptidase domains and a third partial domain and no transmembrane sequence. However, it was later found that this initial report was a sequence error and the correct *Drosophila* CPD sequence encodes three full carboxypeptidase-like domains, a transmembrane domain, and a cytosolic tail [129]. Finally, Lixin Song (working in the author's laboratory) discovered mammalian CPD in a search for an enzyme with CPE-like properties in the *fat* mouse, which lacks CPE activity due to a point mutation (described below in another section). Sequencing of bovine and rat CPD showed that it contained three carboxypeptidase-like domains, a transmembrane domain, and a cytosolic tail [127, 133].

The domain organization of CPD is unique among all other carboxypeptidases in the CPD gene family, which is the M14 metallopeptidase family that includes a total of 25 members di-

Carboxypeptidase D

Discovered in 1995
Rat and bovine carboxypeptidase D (Song and Fricker, JBC, 1995)
Duck homolog – gp180 (Kuroki et al, JBC 1995)
Drosophila homolog – silver gene (Settle et al, PNAS, 1995)

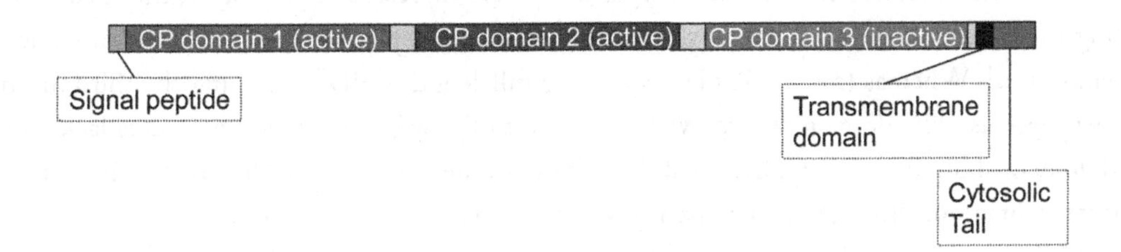

CPD contains 3 carboxypeptidase (CP)-like domains:
CP domains 1 and 2 are enzymatically active, CP domain 3 is inactive

CPD is present in the *trans* Golgi network and also cycles to the cell surface.

FIGURE 17: Key features of carboxypeptidase D (CPD). Simultaneously discovered by three laboratories working in completely different areas, CPD was recognized as an unusual member of the metallocarboxypeptidase gene family due to the presence of multiple carboxypeptidase-like domains followed by a transmembrane domain and cytosolic tail. In species ranging from *Drosophila* to humans, CPD contains 3 carboxypeptidase-like domains, the first two of which are enzymatically active, while the third is inactive due to the absence of critical catalytic and substrate-binding residues. The two active domains have slightly different properties such as pH optima, and the combination of the two domains allows CPD to function in the late Golgi, on the cell surface, and in acidic endosomal compartments after the protein is internalized from the cell surface and routed back to the trans-Golgi network, where it is concentrated.

vided into 4 subfamilies. One distinctive feature of CPD is its membrane attachment through a transmembrane domain. Most other members of the CPD gene family are soluble, and the only other membrane-bound metallocarboxypeptidases are attached through glycosylphosphatidylinositol linkages or amphipathic helixes that peripherally bind to membranes, not through transmembrane domains. The other distinctive feature of CPD is the presence of multiple carboxypeptidase domains in a single protein. Vertebrate CPD contains two functional carboxypeptidase domains followed by a third inactive carboxypeptidase-like domain (Figure 17). This third carboxypeptidase-like domain lacks a critical active site and substrate-binding residues and was therefore predicted to be inactive as a carboxypeptidase; this was subsequently confirmed [134, 135]. However, it remains likely that this third domain has an unidentified function. In a comparison of the amino acid sequences among CPD in different vertebrates, the third domain is as highly conserved as the second carboxypeptidase domain and is more highly conserved than the first carboxypeptidase domain [133]. *Drosophila* CPD also contains two active carboxypeptidase domains followed by a third inactive carboxypeptidase-like domain, although in *Drosophila*, this third domain has little amino acid sequence homology to the comparable domain in vertebrates. A series of transgenic lines of *Drosophila* were created in which different forms of CPD were expressed, including synthetic constructs lacking the third domain or in which the third domain was replaced with one of the other active domains [136]. Whereas transgenic flies expressing full-length CPD containing the third inactive carboxypeptidase-like domain showed viability close to the wild-type flies, constructs lacking the third domain showed greatly reduced viability. These studies suggest that this third domain has an important function, although the precise role of this domain remains unknown.

CPD has a broad pH optimum, reflecting the combined activity of the first domain (which is maximally active at neutral pH) and the second domain (which is maximally active at the acidic pH range of 5–6) [134]. Thus, CPD is able to be active in the trans-Golgi network, on the cell surface, and within the endocytic pathway; these compartments range from neutral to acidic pH. In addition to possessing complementary pH optima, the first and second domains also exhibit different specificity toward C-terminal Lys and Arg residues. The first domain cleaves C-terminal Arg more efficiently than Lys, while the second domain cleaves Lys with higher efficiency [134]. The complementary substrate specificity and pH optima of these two active domains is currently the best hypothesis as to why CPD contains two active carboxypeptidase domains, a feature that has been conserved from *Drosophila* to humans. Interestingly, the nematode *Caenorhabditis elegans* contains a CPD ortholog, based on the presence of multiple carboxypeptidase-like domains, a transmembrane domain, and a cytosolic tail. However, the *C. elegans* CPD ortholog contains only two carboxypeptidase-like domains; the first is predicted to be enzymatically active, while the second lacks many of the necessary residues for catalytic activity and substrate binding and is therefore likely to be inactive. Thus, the fundamental nature of CPD is the presence of at least one active carboxypeptidase domain, one inactive carboxypeptidase-like domain, a transmembrane region, and a cytosolic tail.

The crystal structures of individual carboxypeptidase domains of CPD have been determined and are generally similar to each other. Duck CPD domain 2 was crystallized in 1999 [137], and the active first domain of *Drosophila* CPD was recently crystallized [138]. Both structures show considerable homology to the catalytic region of other members of the M14 zinc-metallocarboxypeptidase family, such as carboxypeptidases A1 and B. However, there are fundamental differences in the residues that interact with the basic charge on the substrate between CPD and carboxypeptidase B, as well as other differences in substrate-binding residues. The catalytic residues are comparable between CPD and all other members of the M14 metallocarboxypeptidase family that have been crystallized. The crystal structure of CPD revealed an additional domain present within each of the carboxypeptidase-like domains; this additional domain has homology to transthyretin and folds into a beta-sheet rich structure. All members of the CPD subfamily (named the N/E subfamily, or M14C family) contain carboxypeptidase-like domains with this additional transthyretin-like domain, which has been proposed to function in protein folding.

3.4 PROHORMONE CONVERTASES 1/3 AND 2

The major peptide-processing endopeptidases are prohormone convertases 1 and 2 (PC1 and PC2) [139]. These enzymes are also referred to as proprotein convertases, and PC1 is also known as PC3 (and usually referred to as PC1/3 to reflect both names). Both PC1/3 and PC2 were originally discovered by John Hutton and colleagues in a search for enzymes that converted proinsulin into insulin [140]. The original names for these enzymes were "type 1" and "type 2" proinsulin-processing endopeptidase [140]. However, due to difficulty in the purification of these enzymes, their molecular characterization required a long and circuitous route involving enzymes found in bacteria (subtilisin), yeast (kex2), and finally mammals (furin). Once furin was identified, two independent groups identified additional mammalian furin-like gene products and these were named PC1 (or 3) and PC2 [141–144]. Subsequently, it was shown that type 1 proinsulin-processing endopeptidase was the same molecular entity as PC1/3, while type 2 proinsulin-processing endopeptidase was PC2 [145, 146]. Because PC1/3 and 2 are in the same gene family as furin, they also show sequence similarity to yeast kex2. PC1/3 and 2 share approximately 50% amino acid sequence similarity in their catalytic domains and also show similar domain organization [141–144]. Both enzymes contain an N-terminal signal peptide domain that targets the protein to the endoplasmic reticulum, a propeptide region that must be removed before the enzyme is fully active, the catalytic domain, a domain referred to as the P-domain that is common to furin and all other members of the PC gene family, and a C-terminal domain (Figure 18). Unlike other members of the PC gene family, the C-terminal regions of PC1/3 and 2 contain are thought to form amphipathic helixes that allow the protein to reversibly attach to membranes as peripheral membrane proteins, in contrast to the intrinsic membrane-binding of furin and CPD. Both PC1/3 and 2 undergo proteolytic processing in their C-terminal region, which activates the enzymes and also reduces their ability to interact with membrane. It should be mentioned that one group of scientists has proposed that PC1/3 and PC2 are attached to membranes through a transmembrane domain, despite the absence of such a hydrophobic domain within the proteins [147, 148]. However, other laboratories have refuted claims that PC1/3 is a transmembrane protein [149], and as neither PC1/3 nor PC2 has the required hydrophobic stretch of amino acids for insertion into the membrane [113], it is highly unlikely that these are integral membrane proteins.

PC1/3 and 2 have similar but not identical distribution patterns [139]. Both are broadly distributed throughout the neuroendocrine system and absent from non-neuroendocrine tissues.

Prohormone Convertase 1 and 2

- 1990- cDNA cloned from homology to furin
- Enzymatic properties identical to "proinsulin-processing endopeptidases" previously found in pancreatic beta cells

Location:

- Neuroendocrine-specific tissue distribution
- Present in peptide-containing secretory vesicles

Activity

- Cleaves after -Arg-Arg— and -Lys-Arg—
- Active at slightly acidic pH (= secretory vesicle environment)
- Activated by Ca^{++} (= secretory vesicle environment)

Elimination causes appropriate effect in biological system

- PC1 knock-out mice are viable but are very small (lack of growth hormones)
- PC2 knock-out mice are generally normal (some peptides are affected)

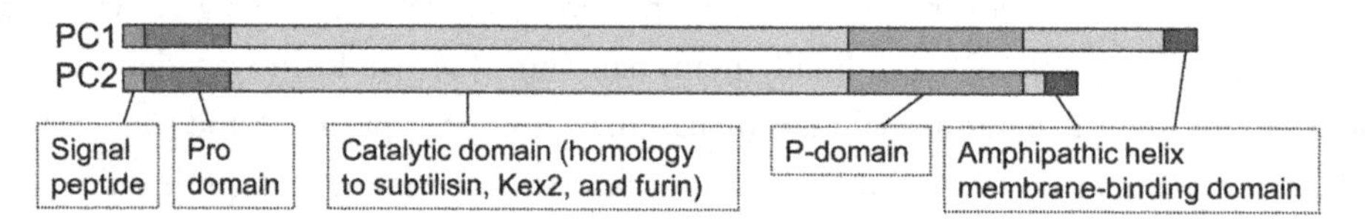

FIGURE 18: Key features of prohormone convertases 1/3 and 2. Both of these endopeptidases are enriched in neuroendocrine tissues such as brain, spinal cord, pituitary, pancreatic islets, adrenal medulla, and intestine. These enzymes function primarily in the secretory granules, cleaving peptide precursors at basic residues. The cleavage sites are usually pairs of basic residues, most often Lys-Arg or Arg-Arg, although the pairs may be separated by 2, 4, or 6 residues. Both PC1/3 and PC2 are activated as the newly formed secretory granules bud from the trans-Golgi network; this activation occurs through a combination of decreasing pH and increasing Ca^{2+} levels. The domain structure of PC1/3 and PC2 has similarities to that of furin, containing a signal peptide domain, a pro domain, a catalytic domain, and a P-domain. Unlike furin, PC1/3 and PC2 are peripheral membrane-bound proteins and contain amphipathic helix domains on their C-termini.

Some neuroendocrine cells express both PC1/3 and 2, while others express just one of the two enzymes. For example, anterior pituitary cells express only PC1/3 and not PC2, while intermediate pituitary cells express both of these enzymes; this difference accounts for the differential processing of proopiomelanocortin into distinct peptides in the anterior and intermediate lobes of the pituitary. Differences in the expression patterns of PC1/3 and 2 are found in brain regions and in other neuroendocrine tissues. Within neuroendocrine cells, PC1/3 and 2 are enriched in the mature secretory granules, where they are maximally active due to their pH optima in the 5–6 range, corresponding to the internal pH of mature secretory granules [139].

Both PC1/3 and 2 cleave substrates at pairs of basic amino acids, usually Lys-Arg or Arg-Arg. In some cases, the pair of basic amino acids is separated by an even number of amino acids [113, 139]. For example, Arg-Xaa-Xaa-Arg is a potential substrate, as is Arg-Xaa-Xaa-Xaa-Xaa-Arg. These sites with non-basic amino acids separating the pair of basic amino acids are often referred to as "single basic sites" or "monobasic sites," although this is technically not correct; there is a pair of basic amino acids present. True single basic sites contain only a Lys or Arg in the cleavage site and do not contain any basic amino acids in the P2, P4, or P6 position. A small number of neuropeptides are cleaved at true single basic sites and some of these appear to be catalyzed by PC1/3 or PC2 [150, 151]. Despite their general similarities, PC1/3 and 2 show subtle preferences for specific sequences. Peptidomic analysis of a large number of peptides detected in the brains of mice lacking either PC1/3 or PC2 activity revealed that PC1/3 has a slight preference for Arg-Arg whereas PC2 prefers Lys-Arg sequences [150, 151]. Of the two enzymes, PC1/3 appears to more readily cleave substrates in which the pairs of basic amino acids are separated by non-basic residues (i.e., Arg-Xaa-Xaa-Arg and longer), although both enzymes can perform this cleavage. PC2 also has greater tolerance for bulky amino acids in the P1′ or P2′ positions [151].

PC1/3 and 2 are activated by Ca^{2+} and acidic pH; these changes occur physiologically when the PCs are packaged into immature granules along with the neuropeptide precursor [113, 139]. It is thought that these are important regulators of PC activity, along with the conversion of the pro form into the active form and proteolytic processing within the C-terminal region.

3.5 CARBOXYPEPTIDASE E

Early studies on proinsulin-processing enzymes identified a carboxypeptidase that produced mature insulin from a processing intermediate [152]. However, this activity was not purified or characterized, aside from demonstration of an acidic pH optimum and sensitivity to chelating agents that bound metal ions. Independently, a metal-dependent carboxypeptidase was discovered in a search for an enzyme that co-localized with enkephalin peptides in bovine adrenal chromaffin granules [153]. This adrenal enzyme produced enkephalin from C-terminally extended processing intermediates and was initially named "enkephalin convertase." Subsequent studies found that the same enzyme is present in adrenal, brain, pituitary, and endocrine pancreas, showing a broad neuroendocrine distribution consistent with a role in the production of many neuropeptides and peptide hormones and not just enkephalins [154]. Therefore, the name carboxypeptidase E (CPE) was considered to be more appropriate than enkephalin convertase [155]. In addition to these two names, the enzyme has also been renamed carboxypeptidase H, although this name is not in common use.

CPE is a zinc metallopeptidase in the same gene family as CPD. Unlike CPD, CPE contains only a single carboxypeptidase domain. Upstream of the carboxypeptidase domain is a short

Carboxypeptidase E

Discovered in 1982
- Named "enkephalin convertase" for initial paper
- Not specific for enkephalin
- Produces most neuropeptides and peptide hormones

Location:
- Neuroendocrine-specific tissue distribution
- Present in peptide-containing secretory granules

Activity
- Cleaves C-terminal Arg and Lys from peptides
- Active at acidic pH (= secretory vesicle environment)

Elimination causes appropriate effect in biological system
- Mice lacking CPE activity have reduced levels of mature peptides and elevated levels of precursors containing C-terminal basic residues

FIGURE 19: Key features of carboxypeptidase E (CPE). CPE was the first mammalian peptide-processing enzyme to be identified. Like PC1/3 and PC2, CPE has a neuroendocrine-specific distribution and is enriched in the mature secretory granules. CPE cleaves C-terminal basic residues (Lys or Arg) from a wide range of peptides, tolerating any residue in the P1 position although with much lower activity toward substrates with Pro in this position. CPE is activated by the decreasing pH of the maturing secretory granules. Mice lacking CPE activity due to a naturally occurring point mutation have altered levels of the vast majority of neuropeptides, with greatly elevated levels of precursors containing C-terminal basic residues and reduced levels of mature forms. The domain structure of CPE includes a signal peptide, a short propeptide domain (that does not need to be removed for the enzyme to be active), a metallocarboxypeptidase domain with homology to carboxypeptidases A and B, an additional beta-sheet rich domain that has structural homology to transthyretin (TT domain) and an amphipathic helix-forming domain that attaches the protein to membranes as a peripheral membrane-binding protein. The TT domain is present in CPD and all other members of the N/E subfamily of carboxypeptidases and is thought to function in protein folding.

propeptide region of 10 amino acid residues (Figure 19). The function of this propeptide region is not known. ProCPE is fully active, with enzymatic properties identical to those of the mature form lacking the propeptide [156]. The propeptide domain does not contribute to the sorting of CPE to the peptide-containing secretory granules [157]. However, the propeptide is 100% conserved among human, monkey, rat, mouse, and other related species and is highly conserved in other mammals; this conservation suggests that the propeptide domain is functional.

The C-terminal region of CPE has a region predicted to form an amphipathic helix, attaching the protein to membranes as a peripheral membrane protein (Figure 19) [158]. It should be mentioned that one group of scientists has claimed that CPE is attached to membranes through a transmembrane domain—despite the absence of an appropriate hydrophobic sequence within the sequence of CPE [159]. However, a transmembrane type of attachment is not compatible with data showing that CPE can be released from membranes at neutral or basic pH values [158, 160, 161]. Differential processing within the C-terminal region of CPE occurs, giving rise to forms that are soluble at acidic pH values. The sites of proteolysis within the C-terminus have not been determined.

CPE is highly regulated by pH [162]. The pH optimum of CPE is in the 5.0–5.5 range, and activity decreases dramatically as the pH is raised outside of this optimal range. This is likely to be of physiological significance, with CPE being nearly inactive at the neutral pH of the endoplasmic reticulum and Golgi, becoming active as the pH drops in the maturing secretory vesicles, and then becoming inactive again when the CPE is released from the cell into the neutral pH extracellular environment.

The substrate specificity of CPE is also consistent with a broad role for this enzyme in the biosynthesis of many different neuropeptides. CPE removes C-terminal Lys and/or Arg residue from a large number of different peptides. The enzyme has no detectable activity for non-basic residues and only weakly cleaves C-terminal His residues [154, 163]. The amino acid in the penultimate position has some influence on enzyme activity; peptides containing the C-terminal sequence Pro-Arg are cleaved orders of magnitude slower than peptides with other amino acids in the penultimate position [164]. This observation fits with the detection in bovine and rat tissues of peptides containing a C-terminal Arg when the penultimate residue is a Pro, as occurs in the peptide alpha-neo-endorphin [165].

The role of CPE in the production of many neuroendocrine peptides was confirmed by analysis of *fat* mice (which are more precisely known as *Cpe*$^{fat/fat}$ mice). These mice lack CPE activity due to a naturally occurring point mutation [91]. In the absence of CPE activity, the levels of the mature forms of many peptides are greatly reduced and there is a dramatic increase in the levels of the peptide processing intermediates containing C-terminal basic residues [128]. This observation fits with the hypothesis that CPE is a major peptide-processing carboxypeptidase.

Attempts to crystallize CPE and determined the structure were not successful, despite numerous attempts using different preparations of enzyme and crystallization methods. CPE is known to be a sticky protein, readily forming aggregates [166], and these aggregates may have interfered with the formation of crystals. The structure of CPE was modeled based on the crystal structures of the homologous CPD [167]. The model of CPE structure shows similar features as CPD, including the substrate-binding residues that are common to CPE and CPD but distinct from members of the A/B family of carboxypeptidases.

3.6 C-TERMINAL AMIDATION

Modifications of the termini serve to protect the peptide from exopeptidases encountered upon secretion from the cell. Peptide stability is a major issue in the endocrine system where the peptide must survive for many minutes in plasma. This presumably explains why a large fraction of peptide hormones are C-terminally amidated. Amidated peptides are also found in the brain but constitute only 10%–20% of the major brain peptides (see Chapter 5), reflecting the reduced importance of stability to exopeptidases for peptides that function in signaling between neurons.

C-terminal amidation is a multistep process involving two catalytic activities (Figure 20). In mammals, both of these activities are produced from the same gene, and collectively the enzyme is known as peptidyl-alpha-amidating monooxygenase (PAM). PAM gene transcripts undergo differential splicing to produce a variety of protein forms [168, 169]. Most of these forms contain an N-terminal enzymatic domain named peptidylglycine alpha-hydroxylating monooxygenase (PHM), a variable length linker domain, and a second enzymatic domain named peptidyl-alpha-hydroxyglycine alpha-amidating lyase (PAL) [170]. Membrane-bound forms of PAM contain a C-terminal transmembrane domain and a cytosolic tail. This cytosolic tail contains some of the same motifs found in furin and CPD, including acidic clusters with phosphorylatable Ser/Thr residues; these motifs function in the trafficking of PAM through the secretory pathway and its return to the trans-Golgi network [171].

PAM recognizes peptides with C-terminal Gly residues and removes the carbon atoms, leaving behind the amine group of the Gly on the C-terminus; this forms the amide (Figure 20). The first step in this process is the oxidation of the Gly by PHM, forming a hydroxyl group on the Gly. PHM is in the same gene family as dopamine beta-hydroxylase and other related enzymes. PHM has 2 copper ions in the active site that participate in the oxidation of the Gly residue. This reaction involves cofactors such as ascorbate and molecular oxygen. Initial studies on PAM focused on soluble forms that contained only the PHM domain and followed the conversion of peptides with C-terminal Gly into the amidated peptide, not the hydroxyl-Gly [172]. Therefore, for this reaction to work, it was necessary to incubate the enzyme and peptide at neutral pH because the second reaction occurred spontaneously only at elevated pH. Although PHM has an acidic pH optimum, at neutral pH PHM is partially active and formed the hydroxyl-Gly, which spontaneously decomposed to form the amidated peptide. The neutral pH optimum of the overall reaction was assumed to represent the pH optimum of the enzyme, as it was not recognized that the second step required an enzyme working at low pH. This presented a problem in the field; the PCs and CPE had acidic pH optima, but it was reported that PAM had a neutral pH optimum. Because PAM must necessarily act following cleavage by PCs and CPE, this was an enigma in the field. Resolution of this problem came from studies that identified a factor that appeared to decrease the pH optimum of the PAM reaction and the surprising finding that this factor was an enzyme encoded by the C-terminal

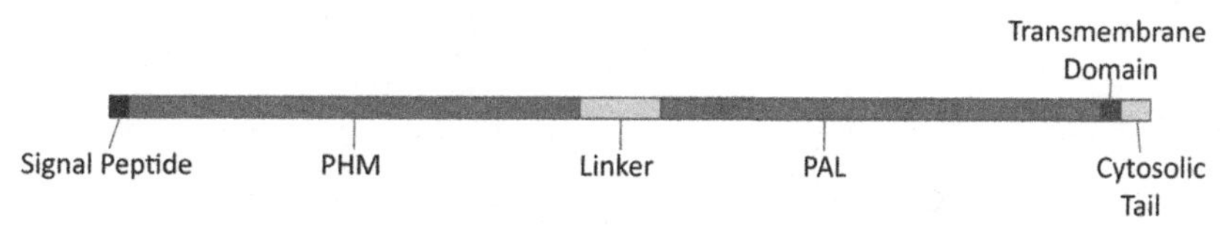

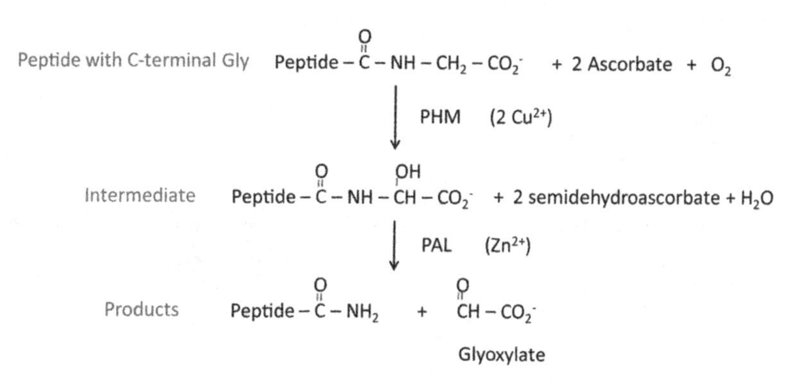

FIGURE 20: Key features of peptide amidation. The formation of a C-terminal amide group on a peptide is a two-step process. Collectively, the two enzymes are referred to as PAM, and in mammals, these enzymes are encoded by the same gene. The gene is alternatively spliced into a range of products, one of which is shown here (in addition, some splice forms contain only the N-terminal enzyme domain). The N-terminal PHM domain catalyzes the oxidation of peptides containing C-terminal Gly residues. Although at neutral or basic pH, this intermediate can spontaneously decompose into the amidated product, at the acidic pH of the mature secretory granule this second step is catalyzed by the enzyme PAL.

region of longer forms of PAM [173]. This second enzyme, PAL, cleaves the hydroxyl Gly residue and allows amidation to proceed in the acidic secretory granule. PAL is a zinc-containing enzyme that has some slight mechanistic similarities to hydrolase such as metallopeptidase, although there are fundamental differences between these two classes of enzymes [174]. The crystal structures of both the PHM and the PAL domains have been determined [170, 174].

3.7 N-TERMINAL MODIFICATIONS OF PEPTIDES

Once the N-terminus of the mature peptide has been liberated, it can be modified by acetylation or pyroglutamylation, although these modifications are relatively rare. The formation of an N-terminal pyroglutamyl residue requires an N-terminal Gln, which is then cyclized into the structure with a blocked N-terminus (Figure 21). Peptides containing an N-terminal Gln are nearly completely found in the pyroglutamate form, indicating that this process is very efficient. It is not clear

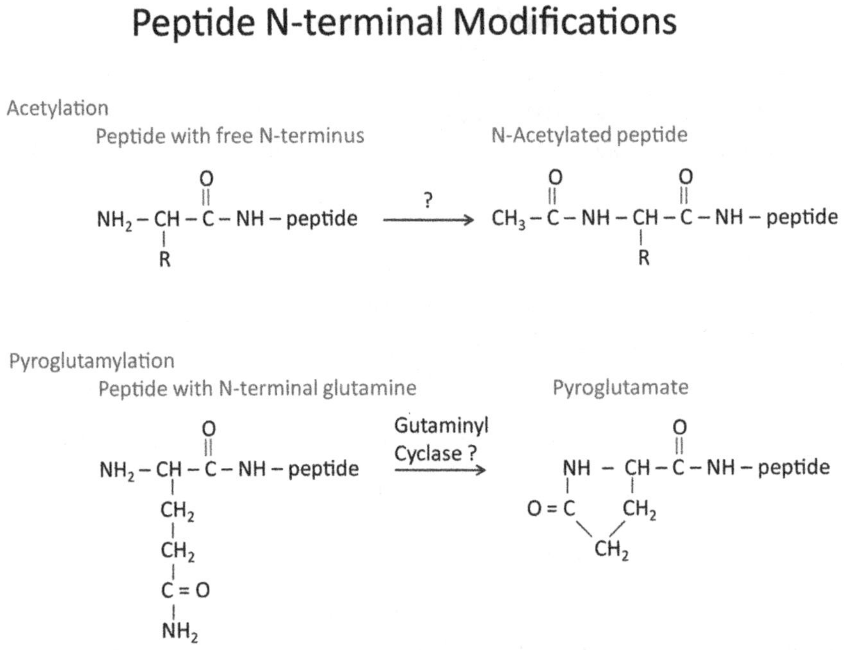

Peptide N-terminal Modifications

Acetylation

Peptide with free N-terminus → (?) → N-Acetylated peptide

$NH_2-CH(R)-C(=O)-NH-peptide \xrightarrow{?} CH_3-C(=O)-NH-CH(R)-C(=O)-NH-peptide$

Pyroglutamylation

Peptide with N-terminal glutamine → (Gutaminyl Cyclase ?) → Pyroglutamate

$NH_2-CH(CH_2-CH_2-C(=O)-NH_2)-C(=O)-NH-peptide \xrightarrow{\text{Gutaminyl Cyclase ?}} NH-CH-C(=O)-NH-peptide$ (ring: NH, $O=C$, CH_2, CH_2)

FIGURE 21: N-terminal modifications of neuropeptides. Acetylation involves the addition of an acetyl group to the N-terminal amine. In the case of alpha-melanocyte-stimulating hormone, a second acetyl group can also be added to the hydroxyl group on the amino acid side chain of an N-terminal Ser residue (not shown). R represents the side chain of the N-terminal residue. The enzyme responsible for N-terminal acetylation has not been identified. Formation of an N-terminal pyroglutamate involves cyclization of an N-terminal glutamine residue, possibly by the enzyme glutaminyl cyclase.

if the cyclization of N-terminal Gln residues proceeds enzymatically or non-enzymatically *in vivo*. Several enzymes have been identified that are capable of converting N-terminal Gln into the corresponding pyroglutamate [175]. One of these enzymes is named glutaminyl-peptide cyclotransferase (commonly referred to as glutaminyl cyclase), and the other is referred to as "isoform of glutaminyl cyclase" or "glutaminyl-peptide cyclotransferase-like" [176]. It was suggested that glutaminyl cyclase is involved in the formation of pyroglutamylated peptides in the brain. However, mice lacking this enzyme due to gene disruption appear to make normal levels of hypothalamic thyrotropin-releasing hormone (a major pyroglutamylated peptide), based on the finding of normal plasma levels of thyroid-stimulating hormone [177]. Thus, it is not clear if glutaminyl cyclase is the major enzyme involved in the production of pyroglutamylated neuropeptides. Some evidence suggests that this enzyme functions in the formation of a pyroglutamate on the N-terminus of an amyloid peptide, producing A-beta pE3-42—which forms toxic aggregates more readily than the full-length unmodified peptide [178]. However, while the formation of this peptide requires pyroglutamylation of an N-terminal glutamic acid residue, and not a glutamine residue, all known neuropeptides with this modification arise from N-terminal glutamine residues. Furthermore, secretory pathway peptides with N-terminal glutamate residues are not found in the pyroglutamyl form. Taken together, these observations raise doubts that that the indentified enzyme functions in the formation of pyroglutamylated neuropeptides.

N-terminal acetylation is relatively rare, having been detected on just a small number of peptides such as alpha-melanocyte-stimulating hormone, beta-endorphin, and joining peptide [179]. All three of these peptides arise from the same precursor protein (proopiomelanocortin, or POMC), suggesting that the acetylating enzyme is specifically expressed in only those cells that express this precursor; otherwise, a larger number of acetylated peptides would likely have been detected in other cell types. The enzyme responsible for N-terminal acetylation has never been conclusively identified; one report in the literature later proved to be incorrect [180]. In addition to N-terminal acetylation, alpha-melanocyte-stimulating hormone is modified by O-acetylation of the N-terminal Ser residue, resulting in the diacetylated peptide. It is assumed that the same enzyme that adds the N-terminal acetyl group is also responsible for the O-acetylation of alpha-melanocyte-stimulating hormone. It is also assumed that the same enzyme that acetylates alpha-melanocyte-stimulating hormone also adds the acetyl group to the N-terminus of beta-endorphin (a Tyr residue) and joining peptide (an Ala residue). However, until the enzyme(s) responsible for neuropeptide acetylation is identified, it remains possible that multiple enzymes are involved in the acetylation of the three different POMC-derived peptides.

3.8 OTHER MODIFICATIONS

The peptide hormone ghrelin, found in stomach endocrine cells, is modified by the addition of an eight carbon acyl group (termed octanoyl) attached to an internal Ser residue [181]. This modification of Ser^3 is required for bioactivity. No other peptide has yet been found to require this same modification. This acylation is catalyzed by the enzyme ghrelin O-acyl-transferase (GOAT), which is co-localized with ghrelin and performs the modification prior to secretion from the cell [182].

It is possible that additional novel modifications occur within neuropeptides. Recent studies using mass spectrometry to detect peptides have detected forms of peptides lacking 18 Da (which corresponds to the loss of water), as well as many forms that cannot be identified using computer programs that search databases of known peptides [32]. These programs are only successful if the modifications are known; novel post-translational modifications will simply prevent peptide identification. Because a large number of the peptides that are detected by mass spectrometry do not match the masses of known peptides, it is possible that these might reflect novel post-translational modifications.

Proteolysis at sites other than the PC recognition motif has also been detected, and this represents a novel post-translational modification. Some of these cleavages occur at single basic residues that lack a second upstream basic residue in the 2, 4, or 6 position, and although these are not consensus sites for the PCs, in some cases, this reaction appears to be catalyzed by PC1/3 or PC2. Other cleavages have been found to occur at non-basic residues, and these are extremely unlikely to be catalyzed by the PCs. For example, neuropeptide FF requires cleavage of an Ala-Phe bond [183]. Either an endopeptidase directly cleaves the precursor at this bond or a PC-like enzyme

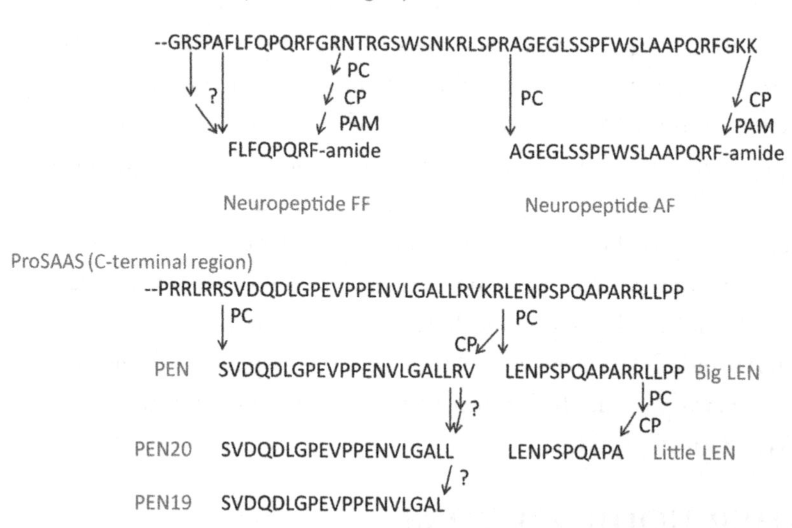

FIGURE 22: Examples of neuropeptide-processing sites that do not contain basic amino acids. A number of peptides that have been identified in the brain or other tissues require processing at non-basic residues. Top: The neuropeptide FF precursor is processed into two amidated peptides. One (neuropeptide AF) requires processing at an Arg-Xaa-Xaa-Arg site on the N-terminus by a PC-like enzyme, trimming of the C-terminus by a CP, and amidation by PAM. (Note that the C-terminus of the precursor ends in basic residues, thus requiring a CP for processing prior to amidation.) The other amidated peptide produced from this region of the precursor is neuropeptide FF. Cleavage on the C-terminal side of neuropeptide FF proceeds via the normal set of enzymes: a PC, a CP, and PAM. However, the N-terminal cleavage site is either an endopeptidase that cleaves the Ala-Phe bond or a PC-like enzyme that cleaves at the upstream basic residue followed by an unknown aminopeptidase trimming to produce the mature neuropeptide FF. The indicated sequence of proneuropeptide FF is bovine. Bottom: Another example of non-basic processing is that of the proSAAS-derived peptide PEN19. This peptide, as well as PEN20 and intact PEN, is abundant in mouse brain. The N-terminus of these PEN peptides is formed by PC-like cleavage at an Arg-Arg site. The C-terminus of PEN19 is either formed by an endopeptidase that cleaves the Leu-Leu bond or from PEN20 by an unknown CP. Also within this C-terminal region of proSAAS are the peptides big and little LEN, which are formed by typical cleavages.

cleaves the precursor at an upstream Arg and then the N-terminus is processed to remove 3 amino acids (Figure 22). Another example of cleavage at non-basic sites is the proSAAS peptide PEN, which is cleaved to produce C-terminally truncated forms, named PEN-20, PEN-19, and others (Figure 22). Some of the C-terminally truncated forms of PEN appear to be present in mouse brain at levels comparable to full-length PEN, suggesting that they are not minor degradation products [184, 185].

Studies examining peptides present in brain extracts detect both the peptides stored within vesicles as well as peptides in other compartments that are exposed to different peptidases. For example, peptides secreted from cells are cleaved by a number of different extracellular peptidases (see Chapter 4). Extracellular breakdown is presumably responsible for some of the non-PC and/or CPE-mediated cleavages that are detected; examples include enkephalin and related peptides lacking the N-terminal Tyr residues, which are known to be produced by extracellular aminopeptidases after secretion. In addition to extracellular processing, some peptide-receptor complexes are internalized, exposing the peptide to peptidases in the endocytic pathway. Finally, there is some evidence that peptide-containing secretory granules are subjected to autophagy rather than secretion, allowing for lysosomal enzymes to cleave the peptides [186].

It has been claimed that lysosomal enzymes such as cathepsin L are involved with the biosynthesis of many peptides in rodent brain, based on the finding that levels of these peptides are greatly reduced in mice lacking cathepsin L activity [187–191]. However, cathepsin L cleaves peptide precursors between the pair of basic residues, generating products with one basic residue on both the N- and C-termini. Thus, if this were a major pathway for the production of enkephalin and other peptides, mice lacking CPE activity would contain precursors with both two C-terminal basic residues (the product of the PCs) and one (the product of cathepsin L). However, the vast majority of the peptide intermediates found in mice lacking CPE activity contain two C-terminal basic residues, not one, and therefore it seems that the major biosynthetic pathway involves PCs and CPE [128, 192]. Still, it is possible that cathepsin L contributes to the processing of peptides and may account for some of the non-PC-mediated cleavages detected in various analyses.

3.9 BIOSYNTHESIS AND SECRETION OF NON-CLASSICAL NEUROPEPTIDES

The preceding sections in this chapter cover the biosynthesis of the classical neuropeptides; those made within the secretory pathway and secreted from cells upon stimulation. The non-classical neuropeptides produced in the cytosol of the cell require completely different enzymes for their production (Figure 23). There are a number of enzymes that are known to cleave cytosolic proteins into peptides. Examples include the proteasome, calpains, and caspases. All of these enzyme pathways are highly regulated, which is a requirement for enzymes that produce non-classical neuropeptides. The proteasome is a large multisubunit protein complex that contains three distinct active subunits,

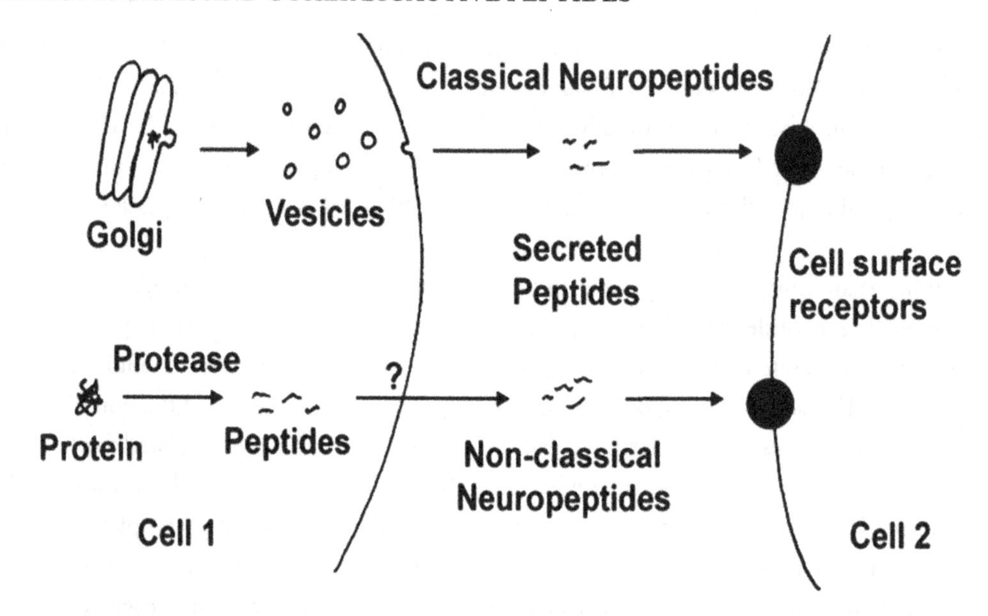

FIGURE 23: Differences between classical neuropeptides and non-classical neuropeptides. Classical neuropeptides are produced by selective cleavage of proteins in the secretory pathway and then secreted when the vesicles fuse with the plasma membrane. In contrast, the putative non-classical neuropeptides are produced from cellular proteins by distinct proteases and then secreted via an unconventional pathway.

named beta-1, beta-2, and beta-5 [193]. These subunits have different specificities, and collectively cleave proteins into peptides of 4–25 amino acids, with an average size of ~10 residues. Recently, the proteasome was shown to contribute to the production of many intracellular peptides, based on the reduction in the levels of these peptides when cells were treated with a proteasome inhibitor [194], but it is not known if the proteasome generates the intracellular peptides that have been proposed to function as non-classical neuropeptides (such as the hemopressins). Calpains are a family of ~15 related enzymes, all of which are activated by Ca^{2+} ions [195]. Once active, the calpains perform limited cleavages of proteins into smaller fragments. Caspases are also a family of related enzymes that are activated by proteolytic conversion of precursor forms into active enzymes, which then cleave proteins at specific sequences [196]. While the calpain cleavage motif is not easily defined, caspases have a clear consensus sequence and always cleave to the C-terminal side of Asp residues. Because the putative non-classical neuropeptides that have been reported do not contain caspase consensus sites on either their N- or C-termini, it is unlikely that the caspases play a role in their production.

In addition to using a different set of enzymes for their biosynthesis, the non-classical neuropeptides also differ from the classical neuropeptides in their secretion mechanism. Whereas the classical neuropeptides are made within secretory granules and are secreted when the vesicles fuse with the membrane, the non-classical neuropeptides produced in the cytosol need to cross the plasma membrane (Figure 23). The mechanism of secretion of non-classical neuropeptides is not known, and there are several possibilities. First, some peptides are able to directly cross membranes without the need for a specific transporting protein [197]. A number of cell-penetrating peptides are known, some containing stretches of hydrophobic residues, while others contain multiple basic residues. It is also possible that specific transporters are involved with the secretion of cytosol-produced peptides. The cellular peptide glutathione is secreted via gap junction hemichannels located on the plasma membrane [198]. Many cells also express P-glycoprotein on the cell surface; this is an ATP-dependent transporter of small molecules that may also transport peptides [199–201]. Basic fibroblast growth factor is produced in the cytosol and secreted via a specific process that involves self-oligomerization, Tec kinase, and heparan sulfate proteoglycans [202]. It is also possible that nonclassical neuropeptides are produced within lysosomes; the contents of certain lysosomes are known to be secreted, and so any peptides produced in these secretory lysosomes would ultimately also be secreted [203, 204]. Exosomes are microparticles released from cells when intracellular multivesicular bodies fuse with the plasma membrane, and represent another potential mechanism for the secretion of peptides [205–208]. Finally, many peptides produced in the cytosol are transported into the lumen of the endoplasmic reticulum by specific proteins named TAP1 and TAP2 (TAP stands for "transporter associated with antigen processing") [209]. Once in the endoplasmic reticulum, the peptides undergo N-terminal processing by a resident aminopeptidase and bind to major histocompatability complex 1 (MHC1). The MHC1/peptide complex is transported through the Golgi to the cell surface, playing an important role in immune system recognition. It thought that peptides that do not bind MHC1 are subsequently degraded, so this route would not be a likely pathway for secretion of cytosol-produced non-classical neuropeptides.

• • • •

CHAPTER 4

Neuropeptides After Secretion: Receptors and Peptidases

Neuropeptides are generally thought to signal nearby cells via activation of cell-surface receptors (Figure 24). Over 100 distinct peptide-binding receptors have been identified and characterized; these include both neuropeptide and peptide hormone receptors. In addition, there are approximately 100 orphan receptors that have no known ligand, and some of these are likely to bind neuropeptides and/or peptide hormones. Peptide receptors are the subject of another book in this series (by Lakshmi Devi and colleagues) and are only briefly described later in this review.

In addition to the well-studied neuropeptide receptors, which clearly mediate many of the biological actions of most neuropeptides, some of the biological effects of peptides occur through other mechanisms. For example, secreted peptides can compete for extracellular peptidases that

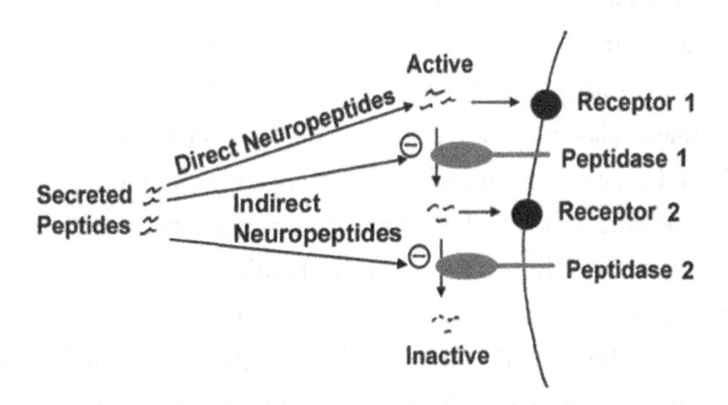

FIGURE 24: Potential mechanisms for peptide-mediated signaling. The conventional view of neuropeptide action is that which is referred to in this figure as "direct neuropeptides." These are peptides that bind to a receptor, thereby influencing the cell's activity. Most of the peptides that are known to be functional neuropeptides fall into this category. Peptides are also cleaved by extracellular peptidases, thus altering their biological properties. In some cases, the modified peptides bind to a second receptor (receptor 2). Eventually, extracellular peptidases degrade the peptide into inactive molecules that do not bind receptors. In some cases, secreted peptides with biological activity were found to function through inhibition of the extracellular peptides and not through binding directly to a receptor; these are termed "indirect neuropeptides" in this figure. Examples of indirect neuropeptides are described in Chapter 5.

normally cleave other endogenous peptides. Extracellular peptidases can either modulate the activity of a peptide by converting it into a form that binds to a distinct receptor or by inactivating the peptide entirely (Figure 24). Inhibition of either type of peptidase will have an effect on the endogenous neuropeptides, thus producing a biological effect. Because this effect does require receptors, albeit in an indirect fashion, an appropriate name for this type of peptide is "indirect neuropeptide" (Figure 24). There are several examples of indirect peptides; these are mentioned in Chapter 5. In this chapter, the focus is on the mechanisms by which secreted peptides can influence other cells through receptor signaling or by interacting with extracellular peptidases.

4.1 NEUROPEPTIDE RECEPTORS

The classic definition of receptor is a protein that binds a chemical messenger and responds by eliciting an intracellular response. In most cases, neuropeptide receptors convert the information from a cell-cell signaling molecule into chemical changes in the cell, causing secondary events that produce a cellular response. In addition to this process, which is termed signal transduction, receptors do two other things: signal amplification and signal processing. Amplification occurs when a neuropeptide binds to a receptor and alters many intracellular molecules by activating an intermediate molecule, termed a second messenger, which then produces further changes such as the opening of an ion channel, activation of an enzyme, turning on gene transcription, or otherwise inducing an intracellular change in many molecules. Signal processing often occurs after receptors are activated, whereby a variety of intracellular events affect the signal and contribute to the response. In some cases, signal processing occurs at the level of the receptor, especially when two or more different neuropeptides bind to the same receptor (described in more detail later in this chapter).

There are four major classes of receptors that bind cell-cell signaling molecules: G-protein-coupled receptors (GPCRs), enzyme-linked receptors, ligand-gated ion channels, and intracellular steroid-type receptors (Figure 25). Of these four groups, neuropeptides are primarily thought to function through binding to GPCRs. Some endocrine peptide hormones also interact with GPCRs, while other peptide hormones bind to enzyme-linked receptors. Examples of this latter group of receptors include those that bind insulin, insulin-like growth factors, and leptin. Some of these receptors are present in the brain, especially the hypothalamus, although technically these are not considered neuropeptide receptors because their ligands are endocrine hormones that are not secreted by neurons.

GPCRs constitute the largest gene family of all receptors, consisting of ~1000 distinct genes in humans and other mammals; this represents 4%–5% of all mammalian genes. The majority of GPCRs are odorant receptors, and of the rest, some are receptors for classical neurotransmitters such as dopamine, serotonin, acetylcholine, and others, while other GPCRs bind to neuropeptides. A number of additional GPCRs have been identified that do not bind known ligands, and some

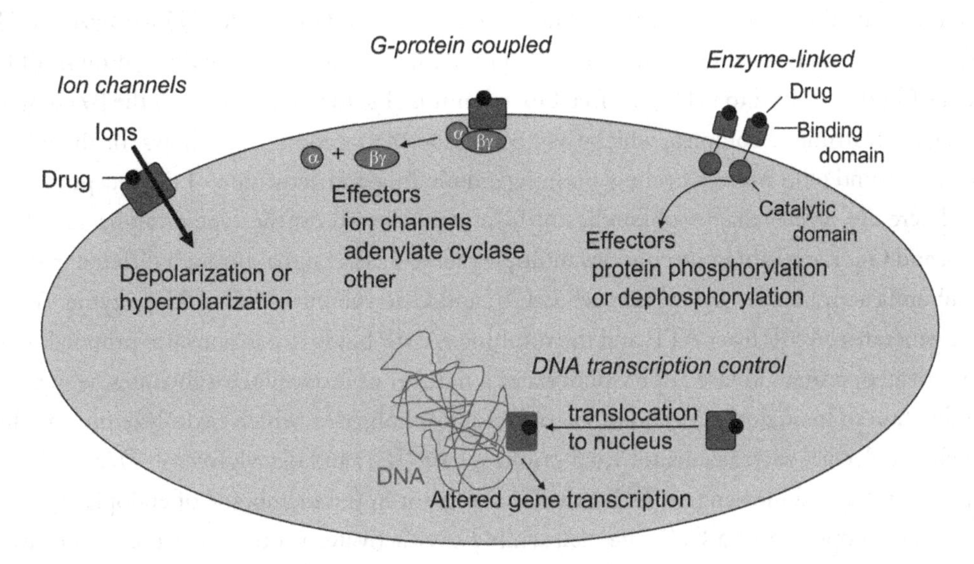

FIGURE 25: Major classes of receptors for chemical messengers. Most neuropeptide receptors are G-protein-coupled. In the endocrine system, peptide hormone receptors are often enzyme-linked receptors, such as the insulin receptor. The other two types of receptors are not known targets for neuropeptides.

of these may bind neuropeptides. It is also possible that receptors for neurotransmitters also bind peptides, thus providing a level of cross-talk and modulation between diverse messengers. This has been demonstrated for the CB1 cannabinoid receptor, which binds endocannabinoids such as anandamide as well as the putative non-classical neuropeptides known as hemopressins (described in more detail in Chapter 5).

GPCRs are integral membrane proteins that contain seven transmembrane spanning domains, and are also referred to as "7-transmembrane spanning receptors" or "heptahelical receptors." They are also known as "metabotropic receptors." Of these various names, GPCR is the most common and is based on the signaling of these receptors through G proteins. Binding of the neuropeptide occurs on the extracellular side of the receptor and leads to changes in the conformation of the receptor protein that are transmitted to the intracellular face of the receptor. The change in receptor conformation affects the association of intracellular G proteins with the receptor. G proteins exist in trimeric protein complexes composed of an α, β, and γ subunit. In the inactive state, the α subunit of the G protein complex is bound to GDP. Upon receptor activation, the G protein releases the GDP molecule, binds a GTP molecule, and the GTP-bound alpha subunit dissociates from the βγ subunits, which remain bound to each other. The free Gα subunit is then able to modulate the activity of effectors such as adenylyl cyclase or phospholipase. In some cases, the βγ complex is also able to modulate the activity of effectors. The α-subunit-bound GTP is slowly hydrolyzed to GDP, and this process can be facilitated by specific cellular proteins that accelerate the hydrolysis of GTP. Once the GTP is cleaved to GDP, the free Gα subunit is able to reassociate with the βγ complex to form the heterotrimeric complex, which then couples to the receptor. This allows for the initiation of the next round of G protein cycling, upon ligand binding and activation of the receptor.

There are several major subfamilies of G proteins, based on the type of alpha subunit: Gs, Gi, Go, and Gq. Each subfamily contains multiple members that signal through different pathways. Gαs subunits activate adenylyl cyclase, while Gαi and Gαo subunits inhibit this enzyme. Adenylyl cyclase generates cAMP from ATP, and the resulting cAMP binds to and activates protein kinase A. Once activated, protein kinase A phosphorylates a number of intracellular substrates, which results in modulation of biological effects. Gαq activates phospholipases, which hydrolyze phospholipids, producing molecules such as inositol 1,4,5-triphosphate (IP3) and diacylglycerol (DAG). Both IP3 and DAG are second messengers; IP3 activates a receptor in the membrane of endoplasmic reticulum leading to opening of a Ca^{2+} channel, while DAG activates protein kinase C and causes the phosphorylation of a variety of molecules. Gα12/13 activates guanine-nucleotide exchange factors, which facilitate the replacement of GDP with GTP within small G proteins such as RhoGTPases. Once activated, the small GTP proteins activate kinases that phosphorylate a variety of target proteins leading to changes in biological effects.

Termination of the activated GPCR occurs through a number of mechanisms. First, the ligand can diffuse away from the receptor, allowing the receptor to revert back to the inactive conformation. Second, once the receptor is activated by ligand binding, the intracellular region of the receptor can be phosphorylated by kinases. Phosphorylated receptor associates with an adaptor protein called arrestin, and the arrestin-bound state of the receptor prevents G proteins from binding and becoming activated; this is termed receptor desensitization. In addition, the arrestin-bound receptor can be removed from the cell surface with the help of endocytic proteins. After endocytosis, the receptor is either recycled back to the membrane or degraded, leading to receptor down-regulation.

Arrestins were originally thought to play a role only in the termination of the signaling through G proteins, but more recently have been found to also participate in the signal transduction. Arrestins form complexes with several signaling proteins such as Src family tyrosine kinases and components of other kinase cascades and transduce the neuropeptide-generated signal to a different set of intracellular molecules than the G proteins. In addition, molecules other than G proteins and arrestins may participate in signal transduction events.

Neuropeptide signaling systems are often more complex than those of conventional neurotransmitters such as acetylcholine in which a single ligand binds to a number of different recep-

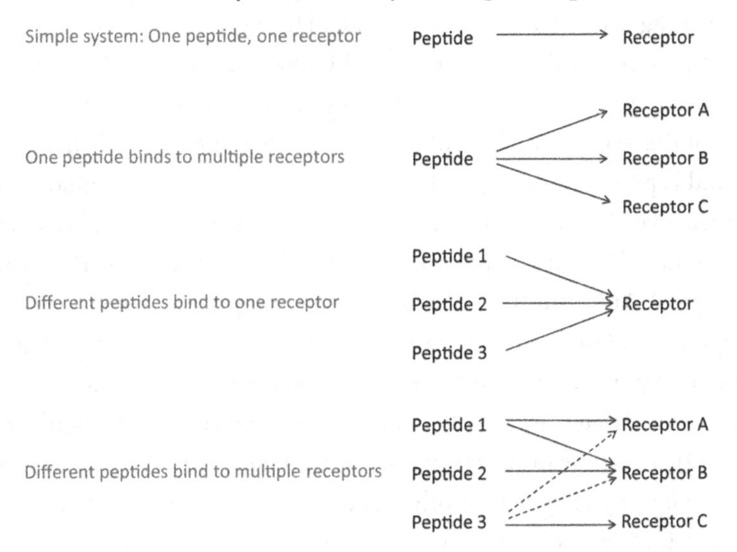

FIGURE 26: Diversity of peptide-receptor signaling. Although there are some examples of the simple system whereby one peptide binds only to one receptor, which does not bind any other molecule, the norm is for much more complexity in the signaling pathways. Often, one peptide can bind to multiple receptors, different peptides can bind to a single receptor, or a combination of these two whereby several peptides are each able to bind to different receptors with varying affinities and/or efficacies.

tors. Most neuropeptides also bind to multiple receptors, but in addition, multiple peptides are usually able to bind to each receptor subtype, adding complexity to the signaling pathways (Figure 26). For example, there are three receptors for opioid peptides; these receptors are named mu, delta, and kappa. Each of these receptors binds endogenous opioid peptides with varying affinity. The mu receptor shows highest affinity for beta-endorphin. The delta receptor has highest affinity for Met- and Leu-enkephalin and slightly lower affinity for larger peptides such as dynorphin A-8. The kappa receptor has highest affinity for longer prodynorphin-derived peptides such as dynorphin A-17 and alpha-neoendorphin, and lower affinity for dynorphin A-8 and shorter peptides. Most other neuropeptide systems are equally complex, although sometimes their names imply more specificity than exists. For example, the oxytocin receptors also bind vasopressin, albeit with lower affinity.

4.2 EXTRACELLULAR PEPTIDASES

A large number of peptidases are present in the extracellular environment where they are either bound to the cell surface of neurons and glia, tethered to the extracellular matrix, or soluble. Most, if not all peptides are cleaved by a variety of peptidases that are present extracellularly. Depending on the peptide and the peptidases, some of these cleavages produce inactive products, thus terminating the receptor-stimulating activity of the neuropeptide. However, for some of the peptides that have been studied, peptidase action does not destroy all biological activity. There are several examples where an extracellular peptidase increases the potency of the peptide toward one particular receptor type. An example of this is the processing of bradykinin, a 9-residue peptide that causes the dilation of blood vessels and is produced in plasma from high-molecular-weight kininogen by the action of kallikrein. The bioactive peptide can be metabolized by several enzymes: angiotensin-converting enzyme, aminopeptidase P, carboxypeptidase M, and carboxypeptidase N. Both carboxypeptidases remove the C-terminal Arg from the 9-residue peptide, forming an 8-residue peptide [9]. While the 9-residue peptide can bind and activate the B2 bradykinin receptor, this peptide is much less active toward the B1 bradykinin receptor than the 8-residue form of the peptide. Thus, the action of carboxypeptidases modulates the activity of bradykinin, converting the peptide from a B2 agonist into a B1 agonist [9]. There are other examples of extracellular processing reactions that shift peptide activity from one receptor type to another, and there are likely to be many additional cases of this phenomenon as further studies investigate this level of regulation.

The major extracellular peptidases are described below. Some of these enzymes are predominantly present in the extracellular space, where their primary function is to process various peptides. Other enzymes listed below are membrane-bound enzymes that are only transiently present on the cell surface and are internalized into the cell from the cell surface where they perform additional functions. These membrane-bound enzymes have their catalytic domains oriented such that they are exposed to the extracellular environment when present on the cell membrane, and when internalized, this catalytic domain is on the inside of vesicles where it can function in the processing of cargo on the way to or from the cell surface. Other extracellular enzymes are primarily cytosolic and are secreted in small amounts by an unknown mechanism. In addition to these peptidases, there are a large number of matrix metalloproteinases and other extracellular enzymes that are thought to function primarily in extracellular protein cleavages, although it is possible that these are also able to cleave small secreted peptides.

4.3 PEPTIDASES PRIMARILY PRESENT IN THE EXTRACELLULAR SPACE

4.3.1 Neprilysin

Neprilysin was originally discovered as a peptide-degrading enzyme present in kidney named neutral endopeptidase [210, 211] and was rediscovered in the brain as an enkephalin-degrading enzyme named "enkephalinase" [6]. This enzyme has also been referred to as membrane metalloendopeptidase, CD10, common acute lymphoblastic leukemia antigen (CALLA), and endopeptidase 24.11 (based on the enzyme commission number EC 3.4.24.11). Because the name neutral endopeptidase is not very specific, the abbreviation of this enzyme (NEP) was modified to "neprilysin," which is now the most commonly used name for this enzyme. Neprilysin is a zinc-dependent metalloprotease that degrades a large number of secreted peptides, such as enkephalins, neuropeptide Y, glucagon, substance P, neurotensin, oxytocin, and bradykinin [212, 213]. Neprilysin shows specificity for cleaving to the N-terminal side of hydrophobic amino acids. Although technically an endopeptidase, meaning that this enzyme cleaves peptides in the middle of the peptide chain and does not require a nearby N- or C-terminus (unlike exopeptidases), neprilysin has a marked preference for cleaving peptides two amino acids from the C-terminus. For this reason, the enzyme was originally referred to as a dipeptidyl carboxypeptidase (more correctly, a peptidyl dipeptidase). However, because neprilysin is able to cleave substrates with a blocked C-terminus, it is technically an endopeptidase. Neprilysin is a type 2 transmembrane protein expressed on the cell surface, with its C-terminal enzymatic domain exposed to the extracellular environment. High levels are present in kidney, especially on the brush border of proximal tubules and on glomerular epithelium. Lower levels are found in many other tissues, including brain where it is predominantly found in the striatum.

Thiorphan and other inhibitors of neprilysin were developed with the idea that they would prevent the degradation of enkephalin, an endogenous opioid peptide, and thus serve as analgesics without addiction potential [214]. Unfortunately, potent inhibitors of neprilysin were not very effective as analgesics. However, one of the side effects of opioid use is constipation (through their action at opioid receptors in the intestine) and inhibition of neprilysin leads to elevated intestinal enkephalin peptides. Racecadotril (also known as acetorphan) is an orally active drug that is metabolized into thiorphan; this drug is now used clinically for treatment of diarrhea [215].

4.3.2 Endothelin-Converting Enzyme (ECE)

ECE is the major enzyme involved in the activation of the endothelins, which are potent vasoconstricting peptides initially secreted from endothelial cells as precursors, called big endothelins. There are three distinct genes for endothelin, and all three genes produce highly homologous peptides that require activation. The conversion of big endothelin into the bioactive form, little endothelin, requires cleavage near the middle of big endothelin at Trp21. This reaction occurs on the cell surface and is catalyzed by ECE, a zinc metalloendopeptidase [216]. ECE is a type 2 membrane protein with the C-terminal catalytic region exposed to the extracellular environment. Several isoforms of ECE have been identified. All of these isoforms are expressed in endothelial cells, polynuclear neutrophils, fibroblasts, cardiomyocytes, and other cell types. In the brain, ECE is broadly expressed at low levels in many neuronal cell types.

With big endothelin 1, 2, or 3 as substrates, ECE is clearly an endopeptidase, cleaving nearly in the middle of the 38-residue peptide [216]. However, with other peptides ECE shows preference for cleavage two residues from the C-terminus, much like neprilysin. Not surprisingly, ECE and neprilysin are in the same gene family, sharing approximately 40% amino acid sequence homology.

4.3.3 Endothelin-Converting Enzyme 2 (ECE2)

Another member of the ECE/neprilysin gene family is ECE2, which has approximately 60% amino acid sequence homology to ECE and 40% sequence homology to neprilysin [217, 218]. The properties of ECE2 are generally similar to ECE; both convert big endothelin into the bioactive forms and also process other peptides [219]. However, ECE2 and ECE have different tissue distributions, cellular distributions, and biochemical properties. ECE is more broadly expressed than ECE2 among tissues. Within cells, both are expressed on the cell surface, but ECE2 may be more involved in processing peptides within endocytic vesicles, based on its presence in these vesicles and its acidic pH optimum. In the brain, ECE2 is most abundant in the dentate gyrus of the hippocampus, whereas ECE is more broadly expressed throughout the brain. ECE2 is not specific for endothelin and is able to process a wide range of peptides in the brain and other tissues.

4.3.4 Angiotensin-Converting Enzyme (ACE)

ACE was originally discovered as the enzyme responsible for the conversion of the inactive angiotensin I into active angiotensin II, a step mediated by removal of the C-terminal dipeptide [220]. ACE is also involved in the degradation of bradykinin and is able to process a number of additional peptides. Additional names for ACE include dipeptidyl carboxypeptidase I, kininase II, and CD143. ACE is found in several forms throughout the body. In most tissues, ACE consists of two distinct enzymatically active domains that exhibit slight differences in enzymatic properties and each of which is capable of functioning independently. The major splice form of the *ACE* gene encodes a type 1 membrane protein in which the N-terminal active domains are extracellular and a single transmembrane domain connects these active enzyme domains to a short cytosolic C-terminal tail. In sperm cells, the gene transcript is spliced to produce a form with only a single enzyme domain. A soluble form of ACE is found in blood; this form is cleaved from the plasma membrane by endopeptidases that separate the enzymatic domains from the transmembrane domain. ACE is highest in the kidney, lung, and intestine, with moderate levels in the brain. Within the brain, levels are highest in choroid plexus and the subfornical organ, with lower levels in the caudate putamen and substantia nigra, and low but detectable levels in many other brain regions [221]. The distribution of ACE in the brain does not correlate with the level of angiotensin, and the role of this enzyme is thought to be general peptide processing rather than a specific involvement in angiotensin activation. Inhibitors of ACE were developed for treatment of hypertension and are still widely used for this purpose [220].

4.3.5 Angiotensin-Converting Enzyme 2 (ACE2)

ACE2 is a member of the same gene family as ACE, but the two enzymes differ in substrate specificity and many other properties [222–224]. Although both are type 1 membrane proteins, ACE2 has only a single catalytic domain (like most enzymes) whereas ACE has two distinct enzymatic domains. ACE2 functions as a carboxypeptidase, removing a single residue (preferably hydrophobic) from the C-terminus of substrates. In the penultimate position, ACE2 prefers a prolyl residue. Thus, a preferred substrate is angiotensin II (C-terminus Pro-Phe) that is converted to the 7-residue peptide named angiotensin 1–7. This peptide does have some bioactivity, although not identical to angiotensin II, and so ACE2 is technically also an "angiotensin-converting enzyme" even though the substrates and products are distinct from those of ACE. In addition to cleaving angiotensin II, ACE2 can also hydrolyze angiotensin 1 into angiotensin 1–9 (with is also inactive), apelin-13 into apelin-12 (both of which are bioactive but with different properties), and des-Arg bradykinin into a product that is inactive toward either bradykinin receptor. Additional substrates are also known, and more are likely to be found.

Like ACE, ACE2 is expressed at high levels in the kidney, with lower levels detectable in the brain and many other tissues [224]. In the brain, ACE2 is highest in hypothalamus. The function of ACE2 in the brain appears to involve regulation of cardiovascular function. Also like ACE, the active domain of ACE2 can be released from the cell surface, leading to soluble forms of the enzyme in the extracellular environment.

4.3.6 Metallocarboxypeptidases

Three members of the CPE zinc metallocarboxypeptidase family (i.e., the M14 family) are primarily present in the extracellular environment: carboxypeptidase A6 (CPA6), carboxypeptidase M (CPM), and carboxypeptidase Z (CPZ) [225–229]. In addition to these three enzymes, other enzymes (such as CPE) are secreted from cells, but in the case of CPE the extracellular environment is not conducive to the enzyme activity (CPE has extremely low activity at $pH > 6.5$). CPM and CPZ are in the same subfamily as CPE, and all three enzymes cleave C-terminal basic residues from peptides. However, both CPM and CPZ are maximally active at neutral pH, and so they are able to cleave within the extracellular environment. CPA6 is in a different subfamily from CPE and the other enzymes. CPA6 cleaves C-terminal hydrophobic residues from many peptides, as well as from proteins, and is also maximally active at neutral pH [230]. CPM is attached to the extracellular membrane through a glycosylphosphatidylinositol anchor, a relatively rare modification that removes the C-terminal ~20–30 residues of the protein and replaces it with the glycosylphosphatidylinositol group [231]. Both CPA6 and CPZ attached to the extracellular matrix; neither of these enzymes has a membrane-binding domain [232, 233]. CPM, CPA6, and CPZ are found in the brain as well as many other tissues. CPA6 and CPZ are expressed in many diverse tissues during development [225, 234].

The precise function of the three extracellular metallocarboxypeptidases is not known. CPM has been shown to convert bradykinin into des-Arg bradykinin, as described above. In the brain, the function of CPM is less clear; the processing of peptide precursors into the mature product requires removal of C-terminal basic residues, but this is thought to be accomplished by CPE (and/or CPD) within the secretory pathway, with no need for processing in the extracellular environment. Therefore, a role for CPM and CPZ in neuropeptide activation is not likely. CPM has been proposed to function in the liberation of Arg, which then serves as a substrate for the production of nitric oxide, an important non-classical neuromodulator. CPZ was proposed to function in the processing of Wnts, specifically those that contain a C-terminal basic residue (such as Wnt4) [235]. Such a function for CPZ has been elegantly demonstrated in two different studies [236, 237]. CPA6 has been found to cleave a number of different substrates, including Leu-enkephalin and angiotensin I, which undergoes two subsequent cleavages by CPA6 to generate the bioactive angiotensin II [230, 232]. Thus, CPA6 can potentially serve as an angiotensin-converting enzyme, performing the same reaction (albeit by two steps) as ACE. Recently, mutations in the *CPA6* gene have been found to be associated with temporal lobe epilepsy, raising the possibility that peptides produced by this enzyme have a neuroprotective effect [238].

4.3.7 Aminopeptidases

A number of extracellular peptide-degrading aminopeptidases have been identified. Of these, the best studied is aminopeptidase N (APN) [239, 240]. This enzyme is also known as alanyl (membrane) aminopeptidase, aminopeptidase M, CD13, EC3.4.11.2, and several other names. APN is a zinc metalloenzyme that preferentially cleaves proteins and peptides with an N-terminal neutral amino acid such as alanine but which is able to cleave a broad range of substrates including Met- and Leu-enkephalin. APN is a type II membrane protein located on the cell surface of many cell types, with the enzymatic domain exposed to the extracellular environment. In the small intestine, APN contributes to the degradation of peptides generated from hydrolysis of proteins in the digestive tract. Human APN is a receptor for coronavirus (hence the name CD13); this property is unrelated to the enzymatic function of APN [241, 242]. In addition to its high expression in intestinal tract and kidney, APN is present in a number of other tissues including liver, lymph node, spleen, and brain. In the brain, APN is thought to represent one of the major extracellular peptide-processing enzymes and inhibitors of this enzyme are able to protect neuropeptides such as enkephalin from degradation.

Another important extracellular aminopeptidase is insulin-regulated aminopeptidase (IRAP), which is also known as placental leucine aminopeptidase, leucyl-cystinyl aminopeptidase, cystinyl aminopeptidase, oxytocinase, EC3.4.11.3, and the angiotensin IV receptor [23, 243]. Although the recommended name is leucyl-cystinyl aminopeptidase, the name IRAP has been widely used in the literature. IRAP is a zinc metalloenzyme in the same gene family as APN. Like APN, IRAP is a type II membrane bound enzyme that is broadly expressed in most tissues. However, unlike APN, the presence of IRAP on the cell surface is regulated. Under basal conditions, IRAP is largely present in intracellular vesicles. Upon stimulation with insulin, IRAP redistributes to the cell surface, along with the insulin-responsive glucose transporter [244]. IRAP cleaves the N-terminal residue of peptide hormones such as Met-enkephalin, dynorphin, oxytocin, vasopressin, and angiotensin III. For many of these peptides, this N-terminal processing step serves to inactive the peptide. Of note, the N-terminal residue of oxytocin and vasopressin is a Cys, which forms a disulfide bond with another Cys in the peptide; therefore, IRAP can cleave a cyclic peptide that is sterically constrained. Mice lacking IRAP due to disruption of the gene show reduced clearance of vasopressin, consistent with the proposed role for IRAP in vasopressin metabolism [245]. In the brain, IRAP is broadly expressed, with highest levels of IRAP mRNA in hippocampus.

IRAP was identified during studies investigating the binding site of angiotensin IV, a biologically active form of angiotensin that facilitates memory retention when injected into the brain of laboratory animals [246]. Angiotensin IV and related peptides were found to bind to a protein, initially named the angiotensin AT(4) receptor. Further analysis revealed that this receptor was identical to IRAP [23, 24, 247]. In addition to binding angiotensin peptides, IRAP also binds forms of hemorphin (LLV-hemorphin 7), a beta-hemoglobin-derived peptide that was previously found to exhibit biological activity. Binding of peptides to IRAP inhibits the catalytic activity, thereby reducing the ability of IRAP to cleave other peptides; this has been proposed as the mechanism of action of angiotensin IV, LLV-hemorphin 7, and other peptides that bind IRAP.

Another aminopeptidase involved in extracellular peptide processing is pyroglutamyl-peptidase II (PPII), which is also known as pyroglutamyl aminopeptidase II, thyrotropin-releasing hormone-degrading ectoenzyme, TRH-degrading ectoenzyme, thyroliberinase, and EC3.4.19.6. PPII is in the same gene family as APN and IRAP. In addition to homology within the catalytic domain, all three of these family members are type II membrane proteins with the zinc-dependent metalloenzyme domain exposed to the extracellular environment. However, PPII is rather specific for thyrotropin releasing hormone (TRH), whereas APN, IRAP, and most of the extracellular peptide processing enzymes described above are able to cleave a large number of peptides. PPII cleaves the N-terminal pyroglutamyl residue from the TRH tripeptide, which inactivates the peptide [248, 249]. The distribution of PPII fits with a predominant, if not exclusive role in the inactivation of TRH; PPII is primarily expressed in the brain and pituitary, with lower levels in lung, liver, and other tissues. In the brain, PPII is specifically found on neuronal cells. Expression of PPII is regulated by drug treatments or physiological conditions that alter TRH, and it is thought that the levels of PPII contribute to the extracellular activity of TRH [250].

The dipeptidyl peptidases (DPPs) are also referred to as dipeptidyl aminopeptidases because they cleave two amino acids from the N-terminus of substrates, releasing a truncated peptide and a dipeptide. DPPs represent a family of ~10 members, although some of these members are not catalytically active (DPP6 and DPP10) due to the absence of a critical Ser residue needed for enzymatic activity. The best studied DPP is DPP4, a type II membrane-bound serine protease located on the cell surface [251]. DPP4 is also known as adenosine deaminase complexing protein-2 and CD26, an antigen involved in T-cell activation. DPP4 preferentially cleaves peptides with Xaa-Pro on the N-terminus, with cleavage following the Pro residue. Substrates include growth factors, chemokines, and bioactive peptides such as glucagon-like peptide-1. DPP4 is found on the cell surface of many cell types, although levels are low in the brain. Because of its role in cleaving glucagon-like peptide-1, inhibitors of DPP4 were developed and show promise as antidiabetic drugs. Because of the low levels of DPP4 in the brain, it is not clear if this enzyme plays a role in the processing of neuropeptides in this tissue.

4.4 OTHER EXTRACELLULAR PROTEASES/PEPTIDASES

In addition to the peptidases described above, there are a large number of other proteolytic enzymes that are either expressed on the cell surface as membrane-bound enzymes or secreted into the extracellular space [212]. These proteases are primarily thought to cleave proteins, often at specific sites, but it is possible that some of these contribute to the cleavage of bioactive peptides. Many of these proteases are assayed with small peptide substrates, indicating that they are capable of cleaving peptides. There are three large families of extracellular proteases: matrix metalloproteinases, tissue kallikreins, and ADAM family proteases.

The matrix metalloproteinases (MMPs) are a large family of over 20 distinct genes, each of which is a zinc-dependent endopeptidase. Some of these enzymes cleave collagen, gelatin, or other extracellular matrix proteins [252]. MMPs play important roles in cell proliferation, migration, and differentiation, as well as other functions. Tissue kallikreins are a family of over a dozen different genes (more in mice and rats than in humans), all of which are serine proteases [253]. While plasma kallikrein functions in the conversion of kininogen into bradykinin, as well as other functions, the primary role of the tissue kallikreins is more complex. Some tissue kallikreins cleave cellular adhesion proteins, while others are associated with neuronal plasticity in the brain. The ADAM family of proteases was named from a feature of the first members of this family, with ADAM standing for the presence of A disintegrin and A metalloprotease domain [254, 255]. ADAMs are multi-domain proteins, although not all of them are active enzymes; some members of this family contain inactive metalloprotease domains that lack critical catalytic residues. Of the group that are active as enzymes, many are involved in the cleavage of membrane-bound proteins, a process termed ectodomain shedding, and these enzymes are referred to as sheddases. Because many of the substrates that are cleaved are cytokines, growth factors, or cell adhesion molecules, the action of the ADAMs serves to regulate cell signaling, cell adhesion, and cell migration [254, 255].

4.5 PEPTIDASES TRANSIENTLY EXPRESSED ON THE CELL SURFACE

In addition to the enzymes described above, which are primarily located on the cell surface (attached to the membrane, in the extracellular matrix, or soluble after secretion), there are other enzymes that are primarily localized to intracellular compartments but which spend a fraction of their time on the cell surface. Even though only transiently expressed on the cell surface, this is sufficient for the enzymes to function in the processing of extracellular proteins and/or peptides. For example, furin and CPD are primarily localized to the trans-Golgi network but move to the cell surface and back to the Golgi. While both of these enzymes have clear functions in the processing of proteins that transit through the secretory pathway (described in Chapter 3), both enzymes also are functional on the cell surface. A number of viral and bacterial proteins are activated by cell surface furin, demonstrating the importance of the cell surface enzyme [112]. Although CPD activity at the cell surface has not been shown to be fundamental to any biological processes, it is likely to be functional because furin and CPD are thought to work as concerted enzymes in the processing; furin initially cleaves proteins at sites containing multiple basic residues, and then CPD trims the basic residues from the C-terminus of the processing intermediate. In addition, cell surface CPD has been found to play a role in the entry of hepatitis B virus into cells [132]. Duck CPD was identified as the cell surface "receptor" that bound duck hepatitis B virus, even though only ~5% of the CPD was actually present on the cell surface at a given time. Therefore, even though both furin and CPD are only transiently present on the cell surface, they are clearly functional in this environment.

Another group of peptidases that have been found to exist outside the cell but which are much more abundant within the cell include a number of cytosolic oligopeptidases. One of these, which is known as oligopeptidase A, thimet oligopeptidase, and endopeptidase 24.15 (EP24.15), was originally described as an enzyme capable of cleaving a number of different neuropeptides [38, 256]. Subsequently, it was found that this enzyme was much more abundant within the cytosol of the cell and is also found in the nucleus. Initially, it was thought that the "secretion" of EP24.15 was the result of cell death, but subsequent studies have shown that this secretion is calcium-regulated and appears to follow an unconventional route, although the precise mechanism is not yet known [257, 258]. Other intracellular cytosolic enzymes capable of cleaving neuropeptides include EP24.16 (also known as neurolysin), insulin-degrading enzyme, and prolyl oligopeptidase [257, 259]. All of these have been found to be secreted from cells and therefore may function in the extracellular environment.

Lysosomal enzymes have also been found to be secreted from cells via a specific process that involves fusion of the lysosomal vesicle with the plasma membrane [203, 204]. Because most lysosomal enzymes are maximally active at acid pH values (~4–5), these enzymes will not be very active in the extracellular environment (~7–7.5). Furthermore, some lysosomal proteases are also unstable at neutral pH, undergoing irreversible inactivation. However, some lysosomal proteases such as cathepsin S have a broader pH range for activity and retain considerable activity at neutral pH [260]. These enzymes may therefore contribute to the extracellular processing of neuropeptides.

• • • •

CHAPTER 5

Representative Neuropeptides

Altogether, over 1000 distinct peptides have been detected in the brain, of which several hundred are derived from proteins present in the secretory pathway [32]. Some of these peptides have been shown to function as neuropeptides, binding to specific receptors and producing a cellular response. Other peptides derived from secretory pathway proteins have been proposed to also function as classical neuropeptides, although further work is required to establish these as *bona fide* neuropeptides. In addition to the peptides derived from secretory pathway proteins, peptides derived from cytosolic or other intracellular proteins can potentially function as non-classical neuropeptides if secreted. Alternatively, these intracellular peptides may play roles in intracellular signaling.

Because the field is still developing, it is not possible to put a precise number on the known neuropeptides; some are well studied and accepted as neuropeptides, while others have met some but not all of the criteria described in Chapter 1. One way of listing the major neuropeptides and/or peptide hormones is to go through the literature and make a note of the number of citations of each peptide, as was done in 1994 by Myers [261]. However, this only reflects the subset of peptides that are well studied and does not reflect the abundance of a particular peptide or its importance. Another way of analyzing peptides is to consider the number of times each peptide has been found in relatively unbiased peptidomics studies (Table 1). As described in Chapter 2, recent approaches to peptide discovery have used mass spectrometry to detect peptides in extracts of mouse brain [32]. Altogether, approximately 250 peptide derived from secretory pathway proteins were detected by this approach (not counting the numerous peptides found in the brain that represent cytosolic protein fragments). While it has been claimed that over 50% of bioactive peptides found in mammals are amidated [262], only 13% of the 250 major brain peptides detected in the peptidomics studies are amidated (i.e., 33 peptides). The discrepancy may be due to three reasons. First, the 13% value from the peptidomics analyses is for all major brain peptides regardless of whether they are functional, and many of these may not be bioactive. Second, in previous studies examining function, those peptides with C-terminal amide groups were initially chosen for further studies because the presence of an amide often reflects bioactive properties. Thus, more of the amidated peptides have been found to be bioactive because they were studied more extensively than other peptides. Third, the "50%" value appears to be based on a survey performed decades ago, and may not be current.

TABLE 1: Analysis of peptides based on peptidomics analysis of mouse brain

PROTEIN	DISTINCT PEPTIDES	TIMES FOUND	RELATIVE ABUNDANCE (%)
ProSAAS	33	1508	17.4
Chromogranin B	22	1126	13.0
Proenkephalin	20	845	9.7
Procholecystokinin	17	681	7.8
Protachykinin A	9	451	5.2
Prodynorphin	9	374	4.3
Proopiomelanocortin	24	359	4.1
Cerebellin 1 precursor protein	4	345	4.0
Secretogranin II	22	326	3.8
Prothyrotropin releasing hormone	13	318	3.7
VGF	12	311	3.6
Chromogranin A	5	216	2.5
Proneurotensin	2	197	2.3
Cerebellin 4	4	183	2.1
Prohormone convertase 2	4	164	1.9
Provasopressin	4	160	1.8
Proneuropeptide Y	2	134	1.5
Secretogranin III	2	132	1.5
Propeptidyl-amidating-monooxygenase	3	131	1.5

TABLE 1: (*continued*)

PROTEIN	DISTINCT PEPTIDES	TIMES FOUND	RELATIVE ABUNDANCE (%)
Promelanin concentrating hormone	3	128	1.5
Pronociceptin/orphanin FQ	4	89	1.0
Protachykinin B	1	77	0.9
Cerebellin 2	3	76	0.9
CART (cocaine- and amphetamine-regulated transcript)	4	75	0.9
Prohormone convertase 1/3	3	57	0.7
Prosomatostatin	2	48	0.6
7B2	3	46	0.5
Provasoactive intestinal peptide	4	34	0.4
Prooxytocin	1	26	0.3
ProPACAP	2	26	0.3
Progastrin releasing peptide	1	17	0.2
Progalanin	2	13	0.1
Proneuromedin B	2	8	0.1

The peptidomics analysis shows that peptides derived from proSAAS are among the most abundant neuropeptides in mouse brain (Table 1). Studies with assays based on antisera to the various proSAAS-derived peptides have confirmed that the levels of these peptides are comparable to those of proenkephalin-derived peptides in mouse brain [263–265]. However, peptides derived from VGF also appear to be as abundant as some of the proSAAS-derived peptides [264], whereas

VGF is 11th on the list based on peptidomics (Table 1). Therefore, the peptidomic needs to be interpreted with caution and relative levels cannot be compared.

Although not biased by preconceived notions about the importance of a particular peptide or the availability of tools such as antisera to detect the peptides, mass spectrometry does have some biases. For a peptide to be detected by mass spectrometry, it must become ionized and enter the gas phase, and the length and amino acid sequence of the peptide greatly influence this. Peptides that are very short (3 amino acids or fewer) or long (30 amino acids or more) are either not detected or show weaker signals than predicted based on the abundance of other peptides derived from the same precursor. For example, the tripeptide thyrotropin-releasing hormone (TRH) has not been detected in peptidomics studies, but larger peptides derived from the TRH precursor have been found a number of times. Because there are five copies of TRH within the precursor, the tripeptide should be present at much higher levels than the other fragments. Another example is Met-enkephalin, a 5-residue peptide. This peptide is detected by mass spectrometry but less frequently than slightly longer peptides derived from the same precursor such as the 7-residue "heptapeptide" and 8-residue "octapeptide." Because there are multiple copies of Met-enkephalin encoded within proenkephalin but only one copy of the heptapeptide and octapeptide, the detection rate does not correlate with the relative abundance of these peptides. Another bias in the peptidomic studies involves the recovery of the peptides during purification steps. Peptides that are highly charged may not bind well to the reverse-phase HPLC columns used in typical peptidomics approaches and therefore will not be detected by the technique. Mass spectrometry of complex mixtures can also present problems when the signals overlap; this occurs when peptides with similar mass to charge ratios co-elute from the HPLC column. As a result, neither of the co-eluting peptides can be identified. Because of these technical limitations, the results of Table 1 should be interpreted loosely; relative levels always need to be quantified by additional techniques such as antiserum-based methods, as has been done for some proSAAS-derived peptides [263, 264], enkephalin [265], and many other neuropeptides.

ProSAAS-derived peptides are the most frequently detected peptides in mouse brain peptidomic studies (Table 1). In addition, peptides derived from proSAAS are also the most numerous with a total of 33 distinct peptides found in studies of wild-type mouse brain. The large number of proSAAS-derived peptides reflects differential processing of the precursor into a variety of products. Most of these processing steps involve cleavage at basic residues, but some represent non-basic cleavage steps and may occur outside the cell in the extracellular space. ProSAAS peptides are described in more detail in a later part of this chapter.

In addition to proSAAS-derived peptides, many peptides derived from chromogranin B (CgB) and proenkephalin are found. Proenkephalin is the precursor of a number of neuropeptides and is described in more detail later in this chapter. CgB is also known as secretogranin 1 and is an abundant secretory granule protein. Unlike typical neuropeptide precursors, which are rapidly converted into peptides within the maturing secretory granule, only a fraction of the CgB and other related granins are converted into peptides within the granule. For this reason, some scientists do not consider the granins to be precursors of neuropeptides. However, some of the CgB-derived peptides have been found to have biological activities [266, 267]. Because the CgB-derived peptides are abundant in the brain, there is no reason to exclude the possibility that they are functional just because much of the precursor remains intact. Peptides derived from other proteins considered to be in the granin family have also been detected in the peptidomics analyses: chromogranin A, secretogranin 2, and VGF (Table 1). As with CgB, peptides derived from each of these other granins have been proposed to function as bioactive peptides in cell-cell signaling, although receptors have not been identified for any of these peptides [266]. Peptides derived from the protein 7B2 are also found in peptidomics analyses (Table 1). 7B2 is a chaperone/inhibitor of prohormone convertase 2 [268]. Although 7B2 is occasionally referred to as a member of the granin family, there are numerous differences between 7B2 and chromogranins [266]. In addition to the granins, other secretory pathway proteins that give rise to peptides found in the peptidomics analyses include the peptide processing enzymes prohormone convertase 1/3, prohormone convertase 2, and propeptidyl-amidating monooxygenase (Table 1). In all cases, the observed peptides are derived from either the N-terminal precursor domain or the C-terminal regions that are known to be cleaved from the active forms of the enzymes. Although considered to be present simply as a result of their removal from the active enzyme, it is conceivable that these N- and/or C-terminal peptides have biological functions as peptides that are independent of their roles in regulating enzyme activity.

Except for the peptides derived from granins and secretory granule enzymes, most of the other peptides found in the peptidomic analyses are either known neuropeptides or are derived from precursors of known neuropeptides. However, one additional group of secretory pathway proteins that are not known to produce biologically active peptides is the cerebellins. These proteins are present in the secretory pathway based on the presence of an N-terminal signal peptide and the known secretion of the protein. Originally discovered in the cerebellum, cerebellin 1 is found throughout the brain, as are cerebellins 2 and 4. The function of the cerebellin proteins is not entirely clear. Recent evidence suggests that cerebellin 1 binds to some isoforms of neurexin and competes with the interaction of neurexin and neuroligin 1 [269]. The potential function of the cleaved cerebellin peptides is unknown.

To complete this book, five of the classical neuropeptide precursors listed in Table 1 and two putative non-classical neuropeptide precursors will be discussed in more detail, in addition to a brief discussion of several examples of indirect neuropeptides. The classical neuropeptide precursors are proneuropeptide Y, the three precursors of opioid peptides (proenkephalin, prodynorphin, and proopiomelanocortin), and proSAAS. The two non-classical precursors chosen for a focus in this chapter are alpha- and beta-hemoglobin, which give rise to a number of peptides with biological activity. Collectively, these systems represent the diversity seen in the various neuropeptide precursors.

5.1 NEUROPEPTIDE Y (NPY)

NPY plays an important role in the regulation of feeding and body weight, as well as a range of other functions [270, 271]. This 36-residue peptide is present in high levels in cells in the arcuate nucleus of the hypothalamus that project to other regions of the brain involved in body weight regulation. NPY was discovered in 1980 during a search for novel amidated peptides [76]. The bioactive pep-

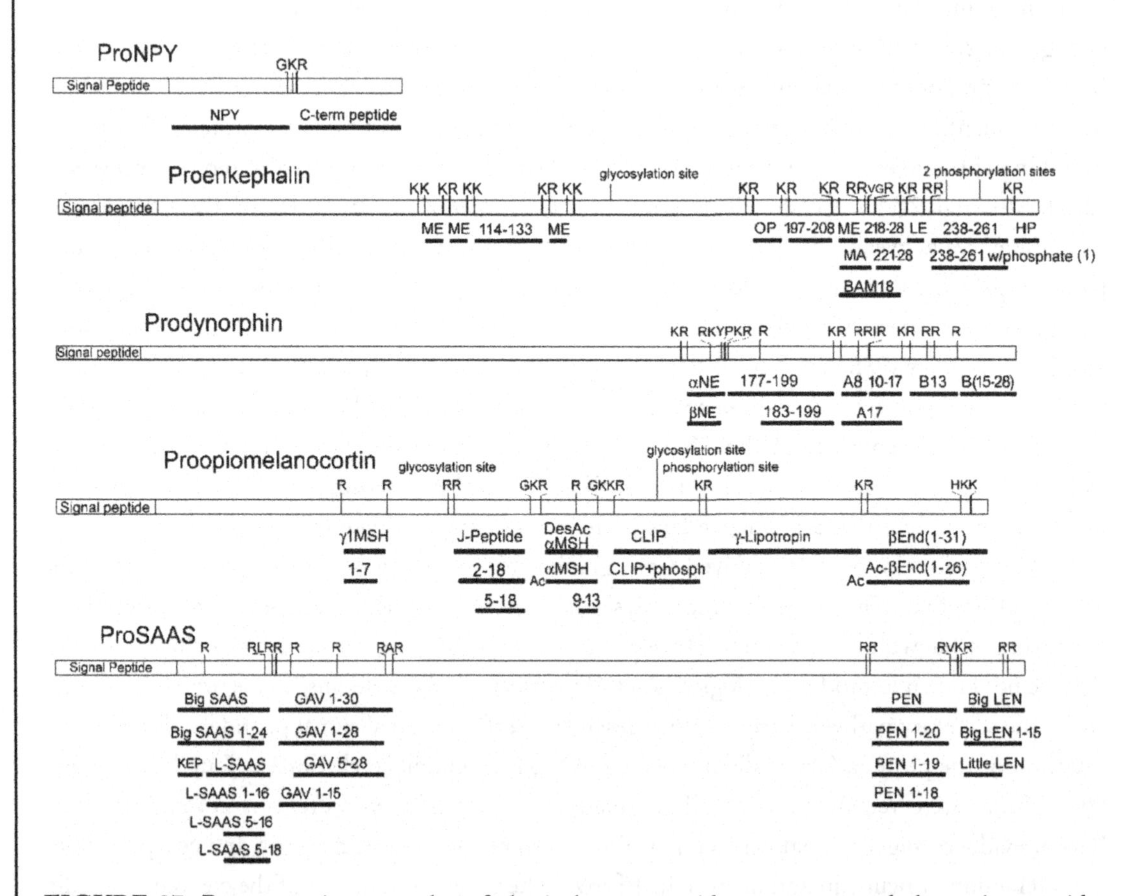

FIGURE 27: Representative examples of classical neuropeptide precursors and the major peptides produced from these precursors. For most of the precursors shown, additional peptides were detected by mass spectrometry, as indicated in Table 1. Some of these minor forms may represent extracellular processing, such as a Leu-enkephalin form, lacking the N-terminal Tyr, which is a known extracellular cleavage product. Abbreviations: NPY, neuropeptide Y; ME, Met-enkephalin; OP, octapeptide; MA, metorphamide; BAM, bovine adrenal medullary peptide; HP, heptapeptide; NE, neoendorphin; A8, dynorphin A8; A17, dynorphin A17; B13, dynorphin B13; B(15–28), dynorphin B residues 15–28; MSH, melanocyte stimulating hormone; J-peptide, joining peptide; CLIP, corticotropin-like intermediate lobe peptide; End, endorphin, L-SAAS, little SAAS. Note that SAAS, GAV, PEN, LEN, and KEP are names, not abbreviations.

tide is generated from proNPY by cleavage at a single processing site that separates the N-terminal NPY sequence from the C-terminal peptide (Figure 27). The single processing site (Gly-Lys-Arg) requires three enzymes to produce the mature form. First, a prohormone convertase cleaves to the C-terminal side of the Arg. Next, the C-terminal Lys and Arg residues are removed by carboxypeptidase E. Finally, the amidating enzyme converts the Gly into an amide group. The mature NPY peptide as well as the C-terminal peptide derived from proNPY are stored within secretory vesicles and secreted upon stimulation. There are five distinct receptors for NPY, all of which are GPCRs [270, 271]. Soon after the finding that injection of NPY into the brain stimulated feeding, there was considerable interest in developing NPY antagonists as drugs to fight obesity. Despite decades of research, this strategy has not been successful, possibly because of adverse effects of NPY antagonists (such as elevated anxiety) due to the multiple roles of NPY in the central nervous system.

5.2 ENDOGENOUS OPIOID PEPTIDES

The endogenous opioid peptides are more complicated than NPY in many respects: the number of biologically active forms, the number of genes, and the role of processing in affecting biological activity. There are two forms of enkephalin: Met-enkephalin and Leu-enkephalin, which were discovered in 1975 during a search for endogenous opioid substances [50]. Met- and Leu-enkephalin are five residue peptides that differ in the C-terminal residue (Tyr-Gly-Gly-Phe-Met/Leu). Both forms of enkephalin have similar biological activities and are agonists of the opioid receptors, with highest affinity for the delta opioid receptor. All of the other peptides known to bind to the opioid receptors share the N-terminal sequence with either Met- or Leu-enkephalin.

Three distinct prohormone precursors give rise to enkephalin-containing peptides: proenkephalin, prodynorphin, and proopiomelanocortin [272–274]. Only proenkephalin is extensively processed into the five residue forms of enkephalins; the others are mainly processed into larger enkephalin-containing peptides such as endorphins and dynorphins. These longer enkephalin-containing peptides bind to delta opioid receptors with lower affinity than Met- or Leu-enkephalin, but the longer peptides bind with higher affinity to mu and/or kappa receptors. Thus, the extent of the processing of each precursor has an effect on the bioactivity of the peptides [275].

Each of the precursors of opioid peptides contains multiple copies of bioactive peptides. Proenkephalin contains one copy of Leu-enkephalin, four copies of Met-enkephalin, and two additional Met-enkephalin-containing peptides named "heptapeptide" and "octapeptide" (Figure 27). Prodynorphin contains three Leu-enkephalin sequences; these are processed into neo-endorphin, dynorphin A, and dynorphin B. Proopiomelanocortin contains one copy of beta-endorphin, an endogenous opioid peptide containing the sequence of Met-enkephalin on the N-terminus. Proopiomelanocortin also encodes several other bioactive peptides such as alpha-melanocyte stimulating hormone and adrenocorticotropic hormone; these peptides are not endogenous opioid peptides and bind to distinct receptors.

The three opioid peptide precursors are expressed in different brain regions [275]. Proenkephalin mRNA is broadly expressed throughout the brain, with highest levels present in the striatum and olfactory bulb. Prodynorphin mRNA is also broadly expressed in the brain and spinal cord. Proopiomelanocortin mRNA is restricted to specific cells in the arcuate region of the hypothalamus, although these cells project to a wide variety of brain regions. All of these opioid peptide precursors are expressed in other neuroendocrine tissues. Proopiomelanocortin is expressed in the anterior and intermediate lobes of the pituitary; in the anterior lobe, it is primarily processed into adrenocorticotropic hormone and full-length beta-endorphin, whereas in the intermediate lobe, it is processed into alpha-melanocyte-stimulating hormone and shorter forms of endorphin. Because the various products of proopiomelanocortin bind to different receptors to elicit distinct effects, a range of physiological changes can be triggered by stimulation of cells expressing this precursor.

5.3 PROSAAS-DERIVED PEPTIDES

In contrast to the examples described above, which are well known and accepted as neuropeptides, peptides derived from proSAAS are not yet well known by most neuroscientists. These peptides are included in this review to illustrate the process required to prove that a peptide is a neuropeptide, and because proSAAS-derived peptides are among the more abundant peptides detected in mouse brain. As shown in Table 1, peptides derived from proSAAS represent approximately 17% of all peptides detected in peptidomics studies on mouse brain. However, this may reflect, to some extent, the large number of peptides produced from this precursor, and studies using antisera-based assays to quantify levels of peptides found that levels of one proSAAS-derived peptide (i.e., PEN) are comparable to levels of enkephalin in mouse brain [263–265].

Altogether, 33 peptides derived from proSAAS have been detected upon peptidomic analysis of mouse brain (Table 1); the major forms are shown in Figure 27. There are four regions of proSAAS that give rise to peptides; these are named SAAS, GAV, PEN, and LEN (the names reflect sequence of amino acids found in each section of the mouse precursor). Most of the peptides shown in Figure 27 reflect cleavage at sites containing basic residues, either pairs (to produce big SAAS, PEN, and either big or little LEN) or single basic amino acids (to produce little SAAS, PEN-20, and some of the other forms). In addition, shorter forms lacking non-basic amino acids are detected (such as PEN 1–19 and PEN 1–18). The enzymes responsible for cleavage at non-basic residues are not known. Some of the cleavages at pairs of basic residues are mediate by PC1/3 and/or PC2, based on the reduction of processed forms observed in mice lacking these processing enzymes [150, 151]. However, pulse-chase analysis indicated that proSAAS was initially cleaved relatively soon after synthesis, suggesting that an ER enzyme such as furin was involved [92]. The cleavage sites needed to generate big SAAS, GAV, PEN, and big LEN are consistent with the consensus site for furin activity, which could explain the early cleavage of proSAAS seen in pulse-chase studies [92].

ProSAAS mRNA is broadly distributed throughout the neuroendocrine system, with detectable levels in virtually every peptide-producing neuroendocrine cell [92, 276]. Because of this, as well as the slightly elevated levels of acidic residues within the proSAAS precursor, some scientists refer to proSAAS as a member of the granin family [266]. However, there is no sequence similarity between proSAAS and any of the granins. Second, granins such as chromogranin A and B, secretogranin II, and VGF are secretory pathway proteins that undergo limited cleavage whereas proSAAS is efficiently cleaved into smaller peptides, as shown in Figure 27. Initial reports on the discovery of proSAAS found evidence that this protein was a potent inhibitor of PC1/3 [92]. For this reason, proSAAS was given the gene name *Pcsk1n*, meaning inhibitor of the *Pcsk1* gene (i.e., PC1/3 protein). The inhibitory sequence within proSAAS is due to the domain between PEN and LEN [277, 278]. The mature forms of these peptides are not PC1/3 inhibitors. Thus, while PC1/3 inhibition may be a function for proSAAS and some of the processing intermediates, this is clearly not a function for the mature forms of the peptides. In addition, although broadly expressed in neuroendocrine cells, proSAAS and its peptides are not evenly distributed. Levels of proSAAS mRNA are much more abundant in the arcuate nucleus of the hypothalamus that other brain regions such as cerebellum or striatum [279]. Also, proSAAS is differentially processed into distinct peptides, much like proenkephalin, prodynorphin, and proopiomelanocortin. For these reasons, it was proposed that proSAAS-derived peptides function as neuropeptides.

To gain a better understanding of the function of proSAAS and its peptides, transgenic mice overexpressing proSAAS and knock-out lacking proSAAS were created; the transgenic mice show elevated body weight, while the knock-out mice are leaner than wild-type littermates [263, 280]. Neither the transgenic mice nor the knock-out mice showed detectable changes in levels of a variety of neuropeptides, suggesting that the effect on body weight was not likely due to altered PC1/3 activity [263, 280]. Using immunofluorescence, the proSAAS peptides PEN and bigLEN were colocalized with NPY in the arcuate nucleus [279]. Collectively, these studies support a role for proSAAS in body weight regulation. To further study this, as well as to identify the specific region of proSAAS responsible for the effects, mice were injected with purified antibodies to different regions of the protein. Antibodies to the peptides PEN and big LEN reduced feeding, whereas antibodies to little LEN and to big and little SAAS were without effect [279]. In electrophysiological assays, the peptide big LEN was able to elicit a rapid response, within seconds of peptide administration, suggesting that it was functioning through a receptor [279]. Taken together, these studies support the hypothesis that proSAAS peptides such as big LEN function in neuropeptides in feeding and body weight regulation. Further proof of this hypothesis requires demonstration of a receptor for big LEN and other proSAAS peptides. Other criteria described in Chapter 1 have already been met (synthesized in neurons, secreted from neurons in a regulated fashion, and knock-out produces a biological effect).

5.4 PEPTIDES THAT FUNCTION AS INDIRECT NEUROPEPTIDES

Considering the large number of peptidases that exist in the extracellular environment and the likely potential that secreted peptides compete for these enzymes, it is surprising that only a few examples exist where the primary function of a secreted peptide is through inhibition of peptidases (Figure 24). Many more interactions probably exist but have not been examined as the general focus of research on bioactive peptides has involved receptor-mediated signaling, and few people consider the possibility that these active peptides exert their effects through interactions with peptidases. Examples of bioactive peptides that appear to function as peptidase inhibitors include angiotensin IV, LVV-hemorphin 7, and opiorphin [24, 281, 282]. The first two of these examples are described in Chapter 4 in the section on IRAP; both of these peptides were found to bind to a protein, named the angiotensin AT(4) receptor, which was subsequently found to be identical to IRAP [24]. Because IRAP is not thought to participate in signal transduction as a true cell surface receptor, it is likely that the biological properties of these two IRAP-binding peptides is due to inhibition of IRAP, which would presumably elevate levels of other neuropeptides in the synapse. The other example, opiorphin, was initially identified as an opioid-like peptide [282]. Further studies found that opiorphin did not bind to any of the opioid receptors but was able to block enkephalin degradation by neprilysin and APN, therefore functioning as an indirect neuropeptide [281]. It is likely that many further examples of indirect neuropeptides will be found as more investigators pursue this possible function for peptides with bioactive properties but which cannot bind directly to neuropeptide receptors.

5.5 NON-CLASSICAL NEUROPEPTIDES

The idea that peptides function as non-classical neurotransmitters analogous to nitric oxide and the lipid-based endocannabinoids is not yet well accepted, but evidence for such a system goes back decades. Early studies on an endogenous valium-like compound identified a peptide, named diazepam-binding inhibitor, which blocked the binding of diazepam to the $GABA_A$ receptor [283]. Subsequently, it was found that this peptide represented a fragment of a cytosolic protein, acyl-CoA-binding protein [284]. Peptides derived from this protein, including several that are nearly identical to the original diazepam-binding inhibitor peptide, were found in peptidomics studies on mouse brain [32]. However, it has not yet been demonstrated that these peptides are secreted from brain cells, which is a necessary requirement to establish them as non-classical neuropeptides.

Recently, a 9-residue peptide derived from alpha-hemoglobin was found to bind to CB1 cannabinoid receptors and function as an antagonist or inverse agonist (i.e., a compound that reduces receptor signaling below endogenous levels) [63]. This peptide, named hemopressin, is produced by an N-terminal cleavage at an Asp-Pro bond. Because Asp-Pro bonds are readily cleaved under acidic conditions and the extraction conditions used to prepare this peptide involved boiling in dilute acid, it is possible that the 9-residue peptide was produced by the extraction procedure and is not an endogenous peptide. Longer forms of the peptide with 2–3 additional N-terminal residues were found in brain extracts prepared without hot acid, and these may therefore represent the true endogenous forms [94]. Interestingly, these N-terminally extended hemopressins also bind to the CB1 cannabinoid receptor but function as agonists. In addition, a homologous peptide from beta-hemoglobin, named VD-beta-hemopressin, was identified and found to also function as an agonist at CB1 receptors [94].

Although alpha- and beta-hemoglobin are not generally considered to be produced in the brain, recent studies have shown that both alpha- and beta-hemoglobin are produced in neurons as well as some non-neuronal cells of the brain and other tissues [285–287]. This fulfills one criterion for non-classical neuropeptides. An additional criterion was met by the finding that the production of the N-terminally extended hemopressins is regulated under specific conditions [20]. Also, RVD-hemopressin can be secreted from brain slices (Gelman and Fricker, unpublished).

Taken together, RVD-hemopressin fits most of the criteria for a non-classical neurotransmitter, which like the lipid-based endocannabinoids is produced on demand and secreted from the cytosol by an undetermined mechanism. Many further studies are needed to identify the mechanism of synthesis and secretion of these peptides. It is important to stress that this concept of non-classical neuropeptides is not well accepted in the field, and it remains to be proven that non-classical neuropeptides are functional within the central nervous system.

• • • •

CHAPTER 6

Concluding Remarks and Future Directions

Peptides represent the largest group of chemical messengers, possibly because the process of generating a new peptide is so simple and occurs naturally through evolution every time a spontaneous mutation arises in a peptide precursor. In a sense, neuropeptides are like words or short sentences; using only 26 letters, it is possible to generate a wide variety of messages. Likewise, with 20 amino acids (not counting modifications), the cell can generate a nearly infinite number of messages. Furthermore, with variable processing of the precursor within the cell, and of the peptide after secretion, even more messages can be created. This all leads to increased complexity. Initial efforts in the field tried to force older simplistic theories on the peptide system, such as the old (and incorrect) dogma that one neuron produces only one neurotransmitter. This dogma was attributed to Henry Dale (the discoverer of oxytocin), although it does not appear in any of his writings and his actual principle was more likely to be that a neuron produces the same neurotransmitters at all of its synapses. Still, the existence of this dogma prevented initial acceptance of the idea that multiple neuropeptides could be produced from the same precursor. History has shown that the simple idea of one neurotransmitter per neuron was incorrect, and that neurons release many molecules for signaling nearby cells.

In addition to trying to force simplicity on complex systems, scientists also try to put things into categories. For example, distinguishing neuropeptides from peptide hormones is difficult. Many of the known peptide hormones also function as neuropeptides, and the distinction between the two categories is not clear-cut. Scientists also have a tendency to think that molecules perform one specific function, but the rule of thumb is that peptides have multiple functions (this is also likely to be true for most proteins). Because most peptides produce a variety of responses depending on the cell type they bind to, it has been difficult to develop specific drugs based on neuropeptides. A relatively small number of drugs that mimic or block the action of peptides have had success as therapeutics, and a much larger number have been found to have side effects that prevented development (although one could argue that this is true for all types of drug targets, not just peptides).

Even after >100 years of research in the field of peptide hormones and neuropeptides, there are still many questions that remain. This review was intended to summarize what is known and well accepted in the field and to also point out some of the emerging ideas and areas for further research. In the field of classical neuropeptides, there are clearly many more surprises waiting as additional neuropeptides are likely to be discovered and new functions found for old peptides. Some putative neuropeptides, such as big LEN and other proSAAS-derived peptides, have met most of the criteria for acceptance as a neuropeptide, requiring only the identification of receptors that bind the peptide. Similarly, the receptor(s) that bind CART peptides have not yet been identified, although evidence suggests that they function through G-protein-coupled receptors [288, 289]. Even well-studied peptide systems such as the enkephalins are likely to yield more surprises. For example, the proenkephalin fragment named bovine adrenal medulla peptide-22 (BAM22) activates a family of sensory neuron-specific G-protein-coupled receptors [290, 291]. The activity of BAM22 at this receptor is not dependent on the opioid receptor-binding region of the peptide. It is possible that many other neuropeptides bind to additional receptors.

Another exciting area in the field of classical neuropeptides is the complexity that results when two different G-protein-coupled receptors form heterodimers. Often, the resulting receptor heterodimer shows a unique pattern of ligand specificity or signals in a different way than each individual receptor [291]. Some G-protein-coupled receptors have recently been found to signal through pathways that do not involve G proteins [292]. Interestingly, the various pathways activated by a receptor are highly dependent on the ligand, with distinct profiles of activation resulting from the binding of different ligands. In simpler terms, receptors are not like a doorbell that produces the same response regardless of the type of finger that pushes the button; instead, receptors are complex signaling entities that respond differently to distinct chemical entities (ligands) to produce unique changes in second messenger pathways. This process is referred to by several names, including biased agonism, ligand bias, and functional selectivity. Much of this has been studied with synthetic ligands, and an exciting future direction will be to test whether different peptide ligands can also produce distinct cellular responses.

The term "non-classical" neuropeptide is relatively new, but the concept that cell-cell signaling molecules are produced from cytosolic proteins is old. The link between non-classical neuropeptides and non-classical neurotransmitters such as nitric oxide is that both types of molecules are made on demand rather than made in advance and stored in vesicles. An appropriate analogy is that of espresso and wine. Classical neuropeptides are like wine—the production takes a long time and cannot be rushed, so the product is stored (in vesicles for neuropeptides, in bottles for wine) and then released from the storage container when needed. In contrast, non-classical neurotransmitters (and espresso) are made when needed because the process of converting the precursor into product is rapid and easily performed. For the latter process, the precursor needs to be present (Arg, in the case of nitric oxide, roasted coffee beans in the case of espresso). Similarly, for non-classical neuropeptides, this made-on-demand process will only work effectively if the precursor protein is present at sufficient levels and the conversion process is rapid. Although the concept of non-classical neuropeptides makes sense and can explain the numerous previous studies that have identified bioactive peptides derived from intracellular proteins (i.e., cytosol-, mitochondrial-, or nuclear-localized proteins), proof of this concept requires demonstration that the peptides are produced in a regulated manner and secreted from cells at levels sufficient to elicit a response in another cell. Typical approaches used to study classical neuropeptides, such as gene knock-out or mRNA knock-down, will be complicated in studies on non-classical neuropeptides because levels of the precursor protein will also be altered by these approaches, thereby complicating the interpretation.

Another emerging idea described in this review is the concept of peptides producing their biological effects indirectly by competing for extracellular processing enzymes, thus altering levels of the direct-acting neuropeptides. As described in this review, some peptides that were previously found to have biological activities attributed to receptor-binding were subsequently found to function instead through inhibition of extracellular enzymes such as the insulin-regulated aminopeptidase or neprilysin [24, 281, 282]. I have proposed the term "indirect neuropeptides" to describe secreted peptides that do not directly bind to receptors but exert their effects through inhibition of peptidases. As with the other emerging ideas, this concept adds to the diversity of cell-cell signaling, giving peptides the ability to influence each other and act in a concerted manner.

Further concepts that are currently emerging, albeit with even less supporting evidence, are those of peptides functioning to alter protein-protein interactions in the extracellular environment and/or inside the cell. A number of cell surface proteins are known to interact with proteins in the extracellular space, either within the extracellular matrix, tethered to adjacent cells, or soluble. The protein-protein interactions cause changes within the cell, such as altered motility, neurite outgrowth, axonal sprouting, and many others. Peptides are able to bind to proteins and either mimic or prevent the binding of other proteins; small peptides have been used by many scientists as research tools to probe protein-protein interactions [27, 28]. However, few people consider it likely that naturally occurring peptides could perform such a role. This concept of peptides interfering with protein-protein interactions is similar to the function of microRNAs and other short non-coding RNAs; previously considered to be junk, these small RNAs are now recognized as important regulators of a large number of gene products [293]. There are many analogies between peptides and small RNAs; both are small oligomers that interact with larger oligomers to alter the function of the larger oligomer. Both also have complex systems for the production and regulation of the small oligomers. Several recent studies have found that intracellular peptides have biological activities [25, 294, 295]. However, much additional work is needed for this concept to be proven. On the other hand, because synthetic peptides are clearly able to affect the activity of many proteins, not just neuropeptide receptors, it seems very likely that nature has exploited this property of peptides to add even more complexity to signaling networks.

References

[1] Strand FL. Neuropeptides: general characteristics and neuropharmaceutical potential in treating CNS disorders. *Prog Drug Res.* 2003;61:1–37.

[2] Zasloff M. Antimicrobial peptides of multicellular organisms. *Nature.* 2002;415:389–95.

[3] Harvey AL. Toxins 'R' Us: more pharmacological tools from nature's superstore. *Trends Pharmacol Sci.* 2002;23:201–3.

[4] Corzo G, Escoubas P. Pharmacologically active spider peptide toxins. *Cell Mol Life Sci.* 2003;60:2409–26.

[5] Goldstein AL, Hannappel E, Kleinman HK. Thymosin beta(4): actin-sequestering protein moonlights to repair injured tissues. *Trends Mol Med.* 2005;11:421–9.

[6] Schwartz JC, de la Baume S, Malfroy B, Patey G, Perdrisot R, Swerts JP, Fournie-Zaluski MC, Gacel G, Roques BP. "Enkephalinase," a newly characterised dipeptidyl carboxypeptidase: properties and possible role in enkephalinergic transmission. *Int J Neurol.* 1980;14:195–204.

[7] Hokfelt T, Broberger C, Xu ZQ, Sergeyev V, Ubink R, Diez M. Neuropeptides—an overview. *Neuropharmacology.* 2000;39:1337–56.

[8] Skidgel RA. Bradykinin-degrading enzymes: structure, function, distribution, and potential roles in cardiovascular pharmacology. *J Cardiovasc Pharmacol.* 1992;20(Suppl 9):S4–9.

[9] Zhang X, Tan F, Zhang Y, Skidgel RA. Carboxypeptidase M and kinin B1 receptors interact to facilitate efficient b1 signaling from B2 agonists. *J Biol Chem.* 2008;283:7994–8004.

[10] Veo K, Reinick C, Liang L, Moser E, Angleson JK, Dores RM. Observations on the ligand selectivity of the melanocortin 2 receptor. *Gen Comp Endocrinol.* 2011;172:3–9.

[11] Eipper BA, Mains RE. Structure and biosynthesis of proACTH/endorphin and related peptides. *Endocrinol Rev.* 1980;1:1–27.

[12] Eipper BA, Mains RE, Herbert E. Peptides in the nervous system. *Trends Neurosci.* 1986;9:463–8.

[13] Zamir N, Weber E, Palkovits M, Brownstein M. Differential processing of prodynorphin and proenkephalin in specific regions of the rat brain. *Proc Natl Acad Sci USA.* 1984;81: 6886–9.

[14] Chavkin C, James IF, Goldstein A. Dynorphin is a specific endogenous ligand of the kappa opioid receptor. *Science.* 1982;215:413–5.

[15] Mansour A, Hoversten MT, Taylor LP, Watson SJ, Akil H. The cloned mu, delta and kappa receptors and their endogenous ligands: evidence for two opioid peptide recognition cores. *Brain Res.* 1995;700:89–98.

[16] Snyder SH, Bredt DS. Nitric oxide as a neuronal messenger. *Trends Pharmacol Sci.* 1991; 12:125–8.

[17] Howlett AC, Breivogel CS, Childers SR, Deadwyler SA, Hampson RE, Porrino LJ. Cannabinoid physiology and pharmacology: 30 years of progress. *Neuropharmacology.* 2004;47(Suppl 1):345–58.

[18] Hatcher NG, Atkins N, Jr, Annangudi SP, Forbes AJ, Kelleher NL, Gillette MU, Sweedler JV. Mass spectrometry-based discovery of circadian peptides. *Proc Natl Acad Sci USA.* 2008;105:12527–32.

[19] Gelman JS, Sironi J, Castro LM, Ferro ES, Fricker LD. Hemopressins and other hemoglobin-derived peptides in mouse brain: comparison between brain, blood, and heart peptidome and regulation in Cpefat/fat mice. *J Neurochem.* 2010;113:871–80.

[20] Gelman JS, Fricker LD. Hemopressin and other bioactive peptides from cytosolic proteins: are these non-classical neuropeptides? *AAPSJ.* 2010;12:279–89.

[21] Boehning D, Snyder SH. Novel neural modulators. *Annu Rev Neurosci.* 2003;26:105–31.

[22] Jaffrey SR, Erdjument-Bromage H, Ferris CD, Tempst P, Snyder SH. Protein S-nitrosylation: a physiological signal for neuronal nitric oxide. *Nat Cell Biol.* 2001;3:193–7.

[23] Vanderheyden PM. From angiotensin IV binding site to AT4 receptor. *Mol Cell Endocrinol.* 2009;302:159–66.

[24] Albiston AL, McDowall SG, Matsacos D, Sim P, Clune E, Mustafa T, Lee J, Mendelsohn FA, Simpson RJ, Connolly LM, Chai SY. Evidence that the angiotensin IV (AT(4)) receptor is the enzyme insulin-regulated aminopeptidase. *J Biol Chem.* 2001;276:48623–6.

[25] Cunha FM, Berti DA, Ferreira ZS, Klitzke CF, Markus RP, Ferro ES. Intracellular peptides as natural regulators of cell signaling. *J Biol Chem.* 2008;283:24448–59.

[26] Arkin MR, Whitty A. The road less traveled: modulating signal transduction enzymes by inhibiting their protein–protein interactions. *Curr Opin Chem Biol.* 2009;13:284–90.

[27] Rubinstein M, Niv MY. Peptidic modulators of protein–protein interactions: progress and challenges in computational design. *Biopolymers.* 2009;91:505–13.

[28] Churchill EN, Qvit N, Mochly-Rosen D. Rationally designed peptide regulators of protein kinase C. *Trends Endocrinol Metab.* 2009;20:25–33.

[29] Armon-Omer A, Levin A, Hayouka Z, Butz K, Hoppe-Seyler F, Loya S, Hizi A, Friedler A, Loyter A. Correlation between shiftide activity and HIV-1 integrase inhibition by a peptide selected from a combinatorial library. *J Mol Biol.* 2008;376:971–82.

[30] Reits E, Griekspoor A, Neijssen J, Groothuis T, Jalink K, van Veelen P, Janssen H, Calafat J, Drijfhout JW, Neefjes J. Peptide diffusion, protection, and degradation in nuclear and cytoplasmic compartments before antigen presentation by MHC class I. *Immunity.* 2003;18:97–108.

[31] Reits E, Neijssen J, Herberts C, Benckhuijsen W, Janssen L, Drijfhout JW, Neefjes J. A major role for TPPII in trimming proteasomal degradation products for MHC class I antigen presentation. *Immunity.* 2004;20:495–506.

[32] Fricker LD. Analysis of mouse brain peptides using mass spectrometry-based peptidomics: implications for novel functions ranging from non-classical neuropeptides to microproteins. *Mol Biosyst.* 2010;6:1355–65.

[33] Svensson M, Skold K, Nilsson A, Falth M, Svenningsson P, Andren PE. Neuropeptidomics: expanding proteomics downwards. *Biochem Soc Trans.* 2007;35:588–93.

[34] Gelman JS, Sironi J, Castro LM, Ferro ES, Fricker LD. Peptidomic analysis of human cell lines. *J Proteome Res.* 2011;10:1583–92.

[35] Bottos A, Rissone A, Bussolino F, Arese M. Neurexins and neuroligins: synapses look out of the nervous system. *Cell Mol Life Sci.* 2011;68:2655–66.

[36] Schweisguth F. Notch signaling activity. *Curr Biol.* 2004;14:R129–38.

[37] Kisselev AF, Akopian TN, Woo KM, Goldberg AL. The sizes of peptides generated from protein by mammalian 26 and 20 S proteasomes. Implications for understanding the degradative mechanism and antigen presentation. *J Biol Chem.* 1999;274:3363–71.

[38] Orlowski M, Reznik S, Ayala J, Pierotti AR. Endopeptidase 24.15 from rat testes. Isolation of the enzyme and its specificity toward synthetic and natural peptides, including enkephalin-containing peptides. *Biochem J.* 1989;261:951–8.

[39] Rioli V, Kato A, Portaro FC, Cury GK, te KK, Vincent B, Checler F, Camargo AC, Glucksman MJ, Roberts JL, Hirose S, Ferro ES. Neuropeptide specificity and inhibition of recombinant isoforms of the endopeptidase 3.4.24.16 family: comparison with the related recombinant endopeptidase 3.4.24.15. *Biochem Biophys Res Commun.* 1998;250:5–11.

[40] Barrett AJ, Brown MA, Dando PM, Knight CG, McKie N, Rawlings ND, Serizawa A. Thimet oligopeptidase and oligopeptidase M or neurolysin. *Methods Enzymol.* 1995;248: 529–56.

[41] Saric T, Graef CI, Goldberg AL. Pathway for degradation of peptides generated by proteasomes: a key role for thimet oligopeptidase and other metallopeptidases. *J Biol Chem.* 2004;279:46723–32.

[42] Gass J, Khosla C. Prolyl endopeptidases. *Cell Mol Life Sci.* 2007;64:345–55.

[43] Grasso G, Rizzarelli E, Spoto G. The proteolytic activity of insulin-degrading enzyme: a mass spectrometry study. *J Mass Spectrom.* 2009;44:735–41.

[44] Berti DA, Morano C, Russo LC, Castro LM, Cunha FM, Zhang X, Sironi J, Klitzke CF, Ferro ES, Fricker LD. Analysis of intracellular substrates and products of thimet oligopeptidase (EC 3.4.24.15) in human embryonic kidney 293 cells. *J Biol Chem.* 2009;284: 14105–16.

[45] Bayliss WM, Starling, E.H. The mechanism of pancreatic secretion. *J Physiol.* 1902;28:325–53.

[46] Mutt V, Jorpes JE, Magnusson S. Structure of porcine secretin. The amino acid sequence. *Eur J Biochem.* 1970;15:513–9.

[47] Vigneaud V, Ressler C, Swan CJ, Roberts CW, Katsoyannis PG, Gordon S. The synthesis of an octapeptide amide with the hormonal activity of oxytocin. *J Am Chem Soc.* 1953;75:4879–80.

[48] Euler US, Gaddum JH. An unidentified depressor substance in certain tissue extracts. *J Physiol.* 1931;72:74–87.

[49] Chang MM, Leeman SE, Niall HD. Amino-acid sequence of substance P. *Nat New Biol.* 1971;232:86–7.

[50] Hughes J, Smith TW, Kosterlitz HW, Fothergill LA, Morgan BA, Morris HR. Identification of two related pentapeptides from the brain with potent opiate agonist activity. *Nature.* 1975;258:577–80.

[51] Goldstein A, Tachibana S, Lowney LI, Hunkapiller M, Hood L. Dynorphin-(1–13), an extraordinarily potent opioid peptide. *Proc Natl Acad Sci USA.* 1979;76:6666–70.

[52] Pert A, Simantov R, Snyder SH. A morphine-like factor in mammalian brain: analgesic activity in rats. *Brain Res.* 1977;136:523–33.

[53] Minamino N, Kangawa K, Fukuda A, Matsuo H, Igarashi M. A new opioid octapeptide related to dynorphin from porcine hypothalamus. *Biochem Biophys Res Commun.* 1980; 95:1475–81.

[54] Kangawa K, Matsuo H. Alpha-Neo-endorphin : a "big" Leu-enkephalin with potent opiate activity from porcine hypothalami. *Biochem Biophys Res Commun.* 1979;86:153–60.

[55] Minamino N, Kangawa K, Chino N, Sakakibara S, Matsuo H. Beta-neo-endorphin, a new hypothalamic "big" Leu-enkephalin of porcine origin: its purification and the complete amino acid sequence. *Biochem Biophys Res Commun.* 1981;99:864–70.

[56] Snyder SH, Childers SR. Opiate receptors and opioid peptides. *Annu Rev Neurosci.* 1979; 2:35–64.

[57] Reinscheid RK, Nothacker H, Bourson A, Ardati A, Henningsen RA, Bunzow JR, Grandy DK, Langen H, Monsma FJ, Civelli O. Orphanin FQ: a neuropeptide that activates an opioidlike G protein-coupled receptor. *Science.* 1995;270:792–4.

[58] Meunier J, Mollereau C, Toll L, Suaudeau C, Moisand C, Alvinerie P, Butour J, Guillemot J, Ferrara P, Monsarrat B, Mazarguil H, Vassart G, Parmentier M, Costentin J. Isolation and structure of the endogenous agonist of opioid receptor-like ORL1 receptor. *Nature.* 1995;377:532–4.

[59] Sakurai T, Amemiya A, Ishii M, Matsuzaki I, Chemelli RM, et al. Orexins and orexin receptors: a family of hypothalamic neuropeptides and G-protein-coupled receptors that regulate feeding behavior. *Cell.* 1998;92:573–85.

[60] Pritchard LE, Armstrong D, Davies N, Oliver RL, Schmitz CA, Brennand JC, Wilkinson GF, White A. Agouti-related protein (83–132) is a competitive antagonist at the human melanocortin-4 receptor: no evidence for differential interactions with pro-opiomelanocortin-derived ligands. *J Endocrinol.* 2004;180:183–91.

[61] Nicolas P, Li CH. Beta-endorphin-(1–27) is a naturally occurring antagonist to etorphine-induced analgesia. *Proc Natl Acad Sci USA.* 1985;82:3178–81.

[62] Hammonds RG, Jr, Nicolas P, Li CH. Beta-endorphin-(1–27) is an antagonist of beta-endorphin analgesia. *Proc Natl Acad Sci USA.* 1984;81:1389–90.

[63] Heimann AS, Gomes I, Dale CS, Pagano RL, Gupta A, de Souza LL, Luchessi AD, Castro LM, Giorgi R, Rioli V, Ferro ES, Devi LA. Hemopressin is an inverse agonist of CB1 cannabinoid receptors. *Proc Natl Acad Sci USA.* 2007;104:20588–93.

[64] Feany MB, Quinn WG. A neuropeptide gene defined by the Drosophila memory mutant amnesiac. *Science.* 1995;268:869–73.

[65] Lin L, Faraco J, Li R, Kadotani H, Rogers W, Lin X, Qiu X, de Jong PJ, Nishino S, Mignot E. The sleep disorder canine narcolepsy is caused by a mutation in the hypocretin (orexin) receptor 2 gene. *Cell.* 1999;98:365–76.

[66] de Lecea L, Kilduff TS, Peyron C, Gao X, Foye PE, Danielson PE, Fukuhara C, Battenberg EL, Gautvik VT, Bartlett FS, Frankel WN, Van Den Pol AN, Bloom FE, Gautvik KM, Sutcliffe JG. The hypocretins: hypothalamus-specific peptides with neuroexcitatory activity. *Proc Natl Acad Sci USA.* 1998;95:322–7.

[67] Hara J, Beuckmann CT, Nambu T, Willie JT, Chemelli RM, Sinton CM, Sugiyama F, Yagami K, Goto K, Yanagisawa M, Sakurai T. Genetic ablation of orexin neurons in mice results in narcolepsy, hypophagia, and obesity. *Neuron.* 2001;30:345–54.

[68] Nishino S, Ripley B, Overeem S, Nevsimalova S, Lammers GJ, Vankova J, Okun M, Rogers W, Brooks S, Mignot E. Low cerebrospinal fluid hypocretin (orexin) and altered energy homeostasis in human narcolepsy. *Ann Neurol.* 2001;50:381–8.

[69] Peyron C, Faraco J, Rogers W, Ripley B, Overeem S, et al. A mutation in a case of early onset narcolepsy and a generalized absence of hypocretin peptides in human narcoleptic brains. *Nat Med.* 2000;6:991–7.

[70] Nishino S, Ripley B, Overeem S, Lammers GJ, Mignot E. Hypocretin (orexin) deficiency in human narcolepsy. *Lancet*. 2000;355:39–40.

[71] Ritchie C, Okuro M, Kanbayashi T, Nishino S. Hypocretin ligand deficiency in narcolepsy: recent basic and clinical insights. *Curr Neurol Neurosci Rep*. 2010;10:180–9.

[72] Zhang Y, Proenca R, Maffei M, Barone M, Leopold L, Friedman JM. Positional cloning of the mouse obese gene and its human homologue. *Nature*. 1994;372:425–32.

[73] Altman J. Weight in the balance. *Neuroendocrinology*. 2002;76:131–6.

[74] Mobbs C, Mizuno T. Leptin regulation of proopiomelanocortin. *Front Horm Res*. 2000; 26:57–70.

[75] Balthasar N, Coppari R, McMinn J, Liu SM, Lee CE, Tang V, Kenny CD, McGovern RA, Chua SC, Jr, Elmquist JK, Lowell BB. Leptin receptor signaling in POMC neurons is required for normal body weight homeostasis. *Neuron*. 2004;42:983–91.

[76] Tatemoto K, Mutt V. Isolation of two novel candidate hormones using a chemical method for finding naturally occurring polypeptides. *Nature*. 1980;285:417–8.

[77] Tatemoto K, Jornvall H, McDonald TJ, Carlquist M, Go VL, Johansson C, Mutt V. Isolation and primary structure of human PHI (peptide HI). *FEBS Lett*. 1984;174:258–61.

[78] Tatemoto K, Jornvall H, Siimesmaa S, Hallden G, Mutt V. Isolation and characterization of cholecystokinin-58 (CCK-58) from porcine brain. *FEBS Lett*. 1984;174:289–93.

[79] Tatemoto K, Rokaeus A, Jornvall H, McDonald TJ, Mutt V. Galanin—a novel biologically active peptide from porcine intestine. *FEBS Lett*. 1983;164:124–8.

[80] Tatemoto K, Carlquist M, Mutt V. Neuropeptide Y—a novel brain peptide with structural similarities to peptide YY and pancreatic polypeptide. *Nature*. 1982;296:659–60.

[81] Stanley BG, Leibowitz SF. Neuropeptide Y injected in the paraventricular hypothalamus: a powerful stimulant of feeding behavior. *Proc Natl Acad Sci USA*. 1985;82:3940–3.

[82] Stanley BG, Chin AS, Leibowitz SF. Feeding and drinking elicited by central injection of neuropeptide Y: evidence for a hypothalamic site(s) of action. *Brain Res Bull*. 1985;14: 521–4.

[83] Amara SG, Jonas V, Rosenfeld MG, Ong ES, Evans RM. Alternative RNA processing in calcitonin gene expression generates mRNAs encoding different polypeptide products. *Nature*. 1982;298:240–4.

[84] Rosenfeld MG, Mermod JJ, Amara SG, Swanson LW, Sawchenko PE, Rivier J, Vale WW, Evans RM. Production of a novel neuropeptide encoded by the calcitonin gene via tissue-specific RNA processing. *Nature*. 1983;304:129–35.

[85] Benarroch EE. CGRP: sensory neuropeptide with multiple neurologic implications. *Neurology*. 2011;77:281–7.

[86] Amara SG, Jonas V, O'Neil JA, Vale W, Rivier J, Roos BA, Evans RM, Rosenfeld MG. Calcitonin COOH-terminal cleavage peptide as a model for identification of novel neuropeptides predicted by recombinant DNA analysis. *J Biol Chem*. 1982;257:2129–32.

[87] Emeson RB, Hedjran F, Yeakley JM, Guise JW, Rosenfeld MG. Alternative production of calcitonin and CGRP mRNA is regulated at the calcitonin-specific splice acceptor. *Nature*. 1989;341:76–80.

[88] Douglass J, McKinzie AA, Couceyro P. PCR differential display identifies a rat brain mRNA that is transcriptionally regulated by cocaine and amphetamine. *J Neurosci*. 1995;15:2471–81.

[89] Hunter RG, Philpot K, Vicentic A, Dominguez G, Hubert GW, Kuhar MJ. CART in feeding and obesity. *Trends Endocrinol Metab*. 2004;15:454–9.

[90] Che FY, Yan L, Li H, Mzhavia N, Devi L, Fricker LD. Identification of peptides from brain and pituitary of Cpe fat /Cpe fat mice *Proc Natl Acad Sci USA*. 2001;98:9971–6.

[91] Naggert JK, Fricker LD, Varlamov O, Nishina PM, Rouille Y, Steiner DF, Carroll RJ, Paigen BJ, Leiter EH. Hyperproinsulinemia in obese fat/fat mice associated with a point mutation in the carboxypeptidase E gene and reduced carboxypeptidase E activity in the pancreatic islets. *Nat Genet*. 1995;10:135–42.

[92] Fricker LD, McKinzie AA, Sun J, Curran E, Qian Y, Yan L, Patterson SD, Courchesne PL, Richards B, Levin N, Mzhavia N, Devi LA, Douglass J. Identification and characterization of proSAAS, a granin-like neuroendocrine peptide precursor that inhibits prohormone processing. *J Neurosci*. 2000;20:639–48.

[93] Sigafoos J, Chestnut WG, Merrill BM, Taylor LCE, Diliberto EJ, Viveros OH. Novel peptides from adrenomedullary chromaffin vesicles. *J Anat*. 1993;183:253–64.

[94] Gomes I, Grushko JS, Golebiewska U, Hoogendoorn S, Gupta A, Heimann AS, Ferro ES, Scarlata S, Fricker LD, Devi LA. Novel endogenous peptide agonists of cannabinoid receptors. *FASEB J*. 2009;23:3020–9.

[95] Sasaki K, Takahashi N, Satoh M, Yamasaki M, Minamino N. A peptidomics strategy for discovering endogenous bioactive peptides. *J Proteome Res*. 2010;9:5047–52.

[96] Che FY, Fricker LD. Quantitative peptidomics of mouse pituitary: comparison of different stable isotopic tags. *J Mass Spectrom*. 2005;40 238–49.

[97] Julka S, Regnier FE. Quantification in proteomics through stable isotope coding: a review. *J Proteome Res*. 2004;3 350–63.

[98] Hsu JL, Huang SY, Chow NH, Chen SH. Stable-isotope dimethyl labeling for quantitative proteomics. *Anal Chem*. 2003;75:6843–52.

[99] Fricker LD, Lim J, Pan H, Che FY. Peptidomics: identification and quantification of endogenous peptides in neuroendocrine tissues. *Mass Spectrom Rev*. 2006;25:327–44.

[100] Ross PL, Huang YN, Marchese JN, Williamson B, Parker K, Hattan S, Khainovski N, Pillai S, Dey S, Daniels S, Purkayastha S, Juhasz P, Martin S, Bartlet-Jones M, He F, Jacobson A, Pappin DJ. Multiplexed protein quantitation in Saccharomyces cerevisiae using amine-reactive isobaric tagging reagents. *Mol Cell Proteomics*. 2004;3:1154–69.

[101] Old WM, Meyer-Arendt K, Aveline-Wolf L, Pierce KG, Mendoza A, Sevinsky JR, Resing KA, Ahn NG. Comparison of label-free methods for quantifying human proteins by shotgun proteomics. *Mol Cell Proteomics*. 2005;4:1487–502.

[102] Tuteja R. Type I signal peptidase: an overview. *Arch Biochem Biophys*. 2005;441:107–11.

[103] Blobel G. Protein targeting. *Biosci Rep*. 2000;20:303–44.

[104] Brodsky JL, Skach WR. Protein folding and quality control in the endoplasmic reticulum: recent lessons from yeast and mammalian cell systems. *Curr Opin Cell Biol*. 2011;23:464–75.

[105] Schwarz F, Aebi M. Mechanisms and principles of N-linked protein glycosylation. *Curr Opin Struct Biol*. 2011;21:576–82.

[106] Beisswanger R, Corbeil D, Vannier C, Thiele C, Dohrmann U, Kellner R, Ashman K, Niehrs C, Huttner WB. Existence of distinct tyrosylprotein sulfotransferase genes: molecular characterization of tyrosylprotein sulfotransferase-2. *Proc Natl Acad Sci USA*. 1998;95:11134–9.

[107] Beinfeld MC. Biosynthesis and processing of pro CCK: recent progress and future challenges. *Life Sci*. 2003;72:747–57.

[108] Bennett HPJ. Glycosylation, phosphorylation, and sulfation of peptide hormones and their precursors. In: Fricker LD, editor. *Peptide Biosynthesis and Processing*. Boca Raton, FL: CRC Press; 1991. pp. 111–40.

[109] Lindberg I, Shaw E. Posttranslational processing of proenkephalin in SK-N-MC cells: evidence for phosphorylation. *J Neurochem*. 1992;58:448–53.

[110] Routledge KE, Gupta V, Balch WE. Emergent properties of proteostasis-COPII coupled systems in human health and disease. *Mol Membr Biol*. 2010;27:385–97.

[111] Thomas G. Furin at the cutting edge: from protein traffic to embryogenesis and disease. *Nat Rev Mol Cell Biol*. 2002;3:753–66.

[112] Nakayama K. Furin: a mammalian subtilisin/Kex2p-like endoprotease involved in processing of a wide variety of precursor proteins. *Biochem J*. 1997;327:625–35.

[113] Seidah NG. The proprotein convertases, 20 years later. *Methods Mol Biol*. 2011;768:23–57.

[114] Roebroek AJ, Schalken JA, Bussemakers MJ, van Heerikhuizen H, Onnekink C, Debruyne FM, Bloemers HP, Van de Ven WJ. Characterization of human c-fes/fps reveals a new transcription unit (fur) in the immediately upstream region of the proto-oncogene. *Mol Biol Rep*. 1986;11:117–25.

[115] Fuller RS, Brake AJ, Thorner J. Intracellular targeting and structural conservation of a prohormone-procesing endoprotease. *Science*. 1989;246:482–6.

[116] Hatsuzawa K, Hosaka M, Nakagawa T, Nagase M, Shoda A, Murakami K, Nakayama K. Structure and expression of mouse furin, a yeast Kex2-related protease. *J Biol Chem*. 1990;265:22075–8.

[117] Molloy SS, Thomas L, VanSlyke JK, Stenberg PE, Thomas G. Intracellular trafficking and activation of the furin proprotein convertase: localization to the TGN and recycling from the cell surface. *EMBO J*. 1994;13:18–23.

[118] Jones BG, Thomas L, Molloy SS, Thulin CD, Fry MD, Walsh KA, Thomas G. Intracellular trafficking of furin is modulated by the phosphorylation state of a casein kinase II site in its cytoplasmic tail. *EMBO J*. 1995;14:5869–83.

[119] Voorhees P, Deignan E, van Donselaar E, Humphrey J, Marks MS, Peters PJ, Bonifacino JS. An acidic sequence within the cytoplasmic domain of furin functions as a determinant of trans-Golgi network localization and internalization from the cell surface. *EMBO J*. 1995;14:4961–75.

[120] Henrich S, Cameron A, Bourenkov GP, Kiefersauer R, Huber R, Lindberg I, Bode W, Than ME. The crystal structure of the proprotein processing proteinase furin explains its stringent specificity. *Nat Struct Biol*. 2003;10:520–6.

[121] Holyoak T, Wilson MA, Fenn TD, Kettner CA, Petsko GA, Fuller RS, Ringe D. 2.4 A resolution crystal structure of the prototypical hormone-processing protease Kex2 in complex with an Ala-Lys-Arg boronic acid inhibitor. *Biochemistry*. 2003;42:6709–18.

[122] Roebroek AJ, Umans L, Pauli IG, Robertson EJ, van Leuven F, Van de Ven WJ, Constam DB. Failure of ventral closure and axial rotation in embryos lacking the proprotein convertase Furin. *Development*. 1998;125:4863–76.

[123] Varlamov O, Fricker LD. Intracellular trafficking of metallocarboxypeptidase D in AtT-20 cells: localization to the trans-Golgi network and recycling from the cell surface. *J Cell Sci*. 1998;111:877–85.

[124] Varlamov O, Kalinina E, Che F, Fricker LD. Protein phosphatase 2A binds to the cytoplasmic tail of carboxypeptidase D and regulates post-TGN trafficking. *J Cell Sci*. 2001;114: 311–22.

[125] Eng FJ, Varlamov O, Fricker LD. Sequences within the cytoplasmic domain of gp180/carboxypeptidase D mediate localization to the trans-Golgi network. *Mol Biol Cell*. 1999;10:35–46.

[126] Song L, Fricker LD. Tissue distribution and characterization of soluble and membrane-bound forms of metallocarboxypeptidase D. *J Biol Chem*. 1996;271:28884–9.

[127] Song L, Fricker LD. Purification and characterization of carboxypeptidase D, a novel carboxypeptidase E-like enzyme, from bovine pituitary. *J Biol Chem.* 1995;270:25007–13.

[128] Zhang X, Che FY, Berezniuk I, Sonmez K, Toll L, Fricker LD. Peptidomics of Cpe(fat/fat) mouse brain regions: implications for neuropeptide processing. *J Neurochem.* 2008;107: 1596–613.

[129] Sidyelyeva G, Fricker LD. Characterization of Drosophila carboxypeptidase D. *J Biol Chem.* 2002;277:49613–20.

[130] Settle SHJ, Green MM, Burtis KC. The silver gene of Drosophila melanogaster encodes multiple carboxypeptidases similar to mammalian prohormone-processing enzymes. *Proc Natl Acad Sci USA.* 1995;92:9470–4.

[131] Ishikawa T, Kuroki K, Lenhoff R, Summers J, Ganem D. Analysis of the binding of a host cell surface glycoprotein to the preS protein of duck hepatitis B virus. *Virology.* 1994; 202:1061–4.

[132] Kuroki K, Eng F, Ishikawa T, Turck C, Harada F, Ganem D. gp180, a host cell glycoprotein that binds duck hepatitis B virus particles, is encoded by a member of the carboxypeptidase gene family. *J Biol Chem.* 1995;270:15022–8.

[133] Xin X, Varlamov O, Day R, Dong W, Bridgett MM, Leiter EH, Fricker LD. Cloning and sequence analysis of cDNA encoding rat carboxypeptidase D. *DNA Cell Biol.* 1997;16: 897–909.

[134] Novikova EG, Eng FJ, Yan L, Qian Y, Fricker LD. Characterization of the enzymatic properties of the first and second domains of metallocarboxypeptidase D. *J Biol Chem.* 1999;274:28887–92.

[135] Eng FJ, Novikova EG, Kuroki K, Ganem D, Fricker LD. gp180, a protein that binds duck hepatitis B virus particles, has metallocarboxypeptidase D-like enzymatic activity. *J Biol Chem.* 1998;273:8382–8.

[136] Sidyelyeva G, Wegener C, Schoenfeld BP, Bell AJ, Baker NE, McBride SM, Fricker LD. Individual carboxypeptidase D domains have both redundant and unique functions in Drosophila development and behavior. *Cell Mol Life Sci.* 2010.

[137] Gomis-Ruth FX, Companys V, Qian Y, Fricker LD, Vendrell J, Aviles FX, Coll M. Crystal structure of avian carboxypeptidase D domain II: a prototype for the regulatory metallocarboxypeptidase subfamily. *EMBO J.* 1999;18:5817–26.

[138] Tanco S, Arolas JL, Guevara T, Lorenzo J, Aviles FX, Gomis-Ruth FX. Structure-function analysis of the short splicing variant carboxypeptidase encoded by Drosophila melanogaster silver. *J Mol Biol.* 2010;401:465–77.

[139] Hoshino AL, Lindberg I. Peptide biosynthesis: prohormone convertases 1/3 and 2. In:

Fricker LDD, LA, editor. *Colloquium Series on Neuropeptides*. Princeton, NJ: Morgan & Claypool Life Sciences; 2012. 108 p.

[140] Davidson HW, Rhodes CJ, Hutton JC. Intraorganellar Ca and pH control proinsulin cleavage in the pancreatic beta-cell via two distinct site-specific endopeptidases. *Nature (Lond)*. 1988;333:93–6.

[141] Smeekens SP, Avruch AS, LaMendola J, Chan SJ, Steiner DF. Identification of a cDNA encoding a second putative prohormone convertase related to PC2 in AtT-20 cells and islets of Langerhans. *Proc Natl Acad Sci USA*. 1991;88:340–4.

[142] Smeekens SP, Steiner DF. Identification of a human insulinoma cDNA encoding a novel mammalian protein structurally related to the yeast dibasic processing protease Kex2. *J Biol Chem*. 1990;265:2997–3000.

[143] Seidah NG, Marcinkiewicz M, Benjannet S, Gaspar L, Beaubien G, Mattei MG, Lazure C, Mbikay M, Chretien M. Cloning and primary sequence of a mouse candidate prohormone convertase PC1 homologous to PC2, furin, and Kex2: distinct chromosomal localization and messenger RNA distribution in brain and pituitary compared to PC2. *Mol Endocrinol*. 1991;5:111–22.

[144] Seidah NG, Gaspar L, Mion P, Marcinkiewicz M, Mbikay M, Chretien M. cDNA sequence of two distinct pituitary proteins homologous to Kex2 and furin gene products: tissue-specific mRNAs encoding candidates for prohormone processing proteinases. *DNA Cell Biol*. 1990;9:415–24.

[145] Bennett DL, Bailyes EM, Nielsen E, Guest PC, Rutherford NG, Arden SD, Hutton JC. Identification of the type 2 proinsulin processing endopeptidase as PC2, a member of the eukaryotic subtilisin family. *J Biol Chem*. 1992;267:15229–36.

[146] Bailyes EM, Shennan KI, Seal AJ, Smeekens SP, Steiner DF, Hutton JC, Docherty K. A member of the eukaryotic subtilisin family (PC3) has the enzymic properties of the type 1 proinsulin-converting endopeptidase. *Biochem J*. 1992;285(Pt 2):391–4.

[147] Arnaoutova I, Smith AM, Coates LC, Sharpe JC, Dhanvantari S, Snell CR, Birch NP, Loh YP. The prohormone processing enzyme PC3 is a lipid raft-associated transmembrane protein. *Biochemistry*. 2003;42:10445–55.

[148] Assadi M, Sharpe JC, Snell C, Loh YP. The C-terminus of prohormone convertase 2 is sufficient and necessary for Raft association and sorting to the regulated secretory pathway. *Biochemistry*. 2004;43:7798–807.

[149] Stettler H, Suri G, Spiess M. Proprotein convertase PC3 is not a transmembrane protein. *Biochemistry*. 2005;44:5339–45.

[150] Wardman JH, Zhang X, Gagnon S, Castro LM, Zhu X, Steiner DF, Day R, Fricker LD.

Analysis of peptides in prohormone convertase 1/3 null mouse brain using quantitative peptidomics. *J Neurochem.* 2010;114:215–25.

[151] Zhang X, Pan H, Peng B, Steiner DF, Pintar JE, Fricker LD. Neuropeptidomic analysis establishes a major role for prohormone convertase-2 in neuropeptide biosynthesis. *J Neurochem.* 2010;112:1168–79.

[152] Zuhlke H, Steiner DF, Lernmark A, Lipsey C. Carboxypeptidase B-like and trypsin-like activities in isolated rat pancreatic islets. *Ciba Found Symp.* 1976;41:183–95.

[153] Fricker LD, Snyder SH. Enkephalin convertase: purification and characterization of a specific enkephalin-synthesizing carboxypeptidase localized to adrenal chromaffin granules. *Proc Natl Acad Sci USA.* 1982;79:3886–90.

[154] Fricker LD, Snyder SH. Purification and characterization of enkephalin convertase, an enkephalin-synthesizing carboxypeptidase. *J Biol Chem.* 1983;258:10950–5.

[155] Fricker LD. Neuropeptide biosynthesis: focus on the carboxypeptidase processing enzyme. *Trends Neurosci.* 1985;8:210–4.

[156] Parkinson D. Two soluble forms of bovine carboxypeptidase H have different NH_2-terminal sequences. *J Biol Chem.* 1990;265:17101–5.

[157] Song L, Fricker LD. The pro region is not required for the expression or intracellular routing of carboxypeptidase E. *Biochem J.* 1997;323:265–71.

[158] Fricker LD, Das B, Angeletti RH. Identification of the pH-dependent membrane anchor of carboxypeptidase E (EC 3.4.17.10). *J Biol Chem.* 1990;265:2476–82.

[159] Dhanvantari S, Arnaoutova I, Snell SR, Steinbach PJ, Hammond K, Caputo GA, London E, Loh YP. Carboxypeptidase E, a prohormone sorting receptor, is anchored to secretory granules via a C-terminal transmembrane insertion. *Biochemistry.* 2002;41:52–60.

[160] Mitra A, Song L, Fricker LD. The C-terminal region of carboxypeptidase E is involved in membrane binding and intracellular routing in AtT-20 cells. *J Biol Chem.* 1994;269:19876–81.

[161] Varlamov O, Fricker LD. The C-terminal region of carboxypeptidase E involved in membrane binding is distinct from the region involved with intracellular routing. *J Biol Chem.* 1996;271:6077–83.

[162] Greene D, Das B, Fricker LD. Regulation of carboxypeptidase E: effect of pH, temperature, and Co++ on kinetic parameters of substrate hydrolysis. *Biochem J.* 1992;285:613–8.

[163] Smyth DG, Maruthainar K, Darby NJ, Fricker LD. C-terminal processing of neuropeptides: involvement of carboxypeptidase H. *J Neurochem.* 1989;53:489–93.

[164] Chen H, Jawahar S, Qian Y, Duong Q, Chan G, Parker A, Meyer JM, Moore KJ, Chayen S, Gross DJ, Glasser B, Permutt MA, Fricker LD. A missense polymorphism in the hu-

man carboxypeptidase E gene alters its enzymatic activity: possible implications in type 2 diabetes mellitus. *Hum Mutat.* 2001;18:120–31.

[165] Kangawa K, Minamino N, Chino N, Sakakibara S, Matsuo H. The complete amino acid sequence of alpha-neo-endorphin. *Biochem Biophys Res Commun.* 1981;99:871–8.

[166] Song L, Fricker LD. Calcium- and pH-dependent aggregation of carboxypeptidase E. *J Biol Chem.* 1995;270:7963–7.

[167] Aloy P, Companys V, Vendrell J, Aviles FX, Fricker LD, Coll M, Gomis-Ruth FX. The crystal structure of the inhibitor-complexed carboxypeptidase D domain II as a basis for the modelling of regulatory carboxypeptidases. *J Biol Chem.* 2001;276:16177–84.

[168] Ouafik LH, Stoffers DA, Campbell TA, Johnson RC, Bloomquist BT, Mains RE, Eipper BA. The multifunctional peptidylglycine alpha-amidating monooxygenase gene: exon/intron organization of catalytic, processing, and routing domains. *Mol Endocrinol.* 1992;6: 1571–84.

[169] Eipper BA, Milgram SL, Husten EJ, Yun HY, Mains RE. Peptidylglycine alpha-amidating monooxygenase: a multifunctional protein with catalytic, processing, and routing domains. *Protein Sci.* 1993;2:489–97.

[170] Prigge ST, Mains RE, Eipper BA, Amzel LM. New insights into copper monooxygenases and peptide amidation: structure, mechanism and function. *Cell Mol Life Sci.* 2000;57: 1236–59.

[171] Milgram SL, Mains RE, Eipper BA. Identification of routing determinants in the cytoplasmic domain of a secretory granule-associated integral membrane protein. *J Biol Chem.* 1996;271:17526–35.

[172] Eipper BA, Mains RE. Peptide alpha-amidation. *Annu Rev Physiol.* 1988;50:333–44.

[173] Eipper BA, Perkins SN, Husten EJ, Johnson RC, Keutmann HT, Mains RE. Peptidyl-α-hydroxyglycine α-amidating lyase: purification, characterization, and expression. *J Biol Chem.* 1991;266:7827–33.

[174] Chufan EE, De M, Eipper BA, Mains RE, Amzel LM. Amidation of bioactive peptides: the structure of the lyase domain of the amidating enzyme. *Structure.* 2009;17:965–73.

[175] Pohl T, Zimmer M, Mugele K, Spiess J. Primary structure and functional expression of a glutaminyl cyclase. *Proc Natl Acad Sci USA.* 1991;88:10059–63.

[176] Cynis H, Rahfeld JU, Stephan A, Kehlen A, Koch B, Wermann M, Demuth HU, Schilling S. Isolation of an isoenzyme of human glutaminyl cyclase: retention in the Golgi complex suggests involvement in the protein maturation machinery. *J Mol Biol.* 2008;379:966–80.

[177] Schilling S, Kohlmann S, Bauscher C, Sedlmeier R, Koch B, Eichentopf R, Becker A, Cynis H, Hoffmann T, Berg S, Freyse EJ, von Horsten S, Rossner S, Graubner S, Demuth HU. Glutaminyl cyclase knock-out mice exhibit slight hypothyroidism but no

hypogonadism: implications for enzyme function and drug development. *J Biol Chem.* 2011;286:14199–208.

[178] Hartlage-Rubsamen M, Morawski M, Waniek A, Jager C, Zeitschel U, Koch B, Cynis H, Schilling S, Schliebs R, Demuth HU, Rossner S. Glutaminyl cyclase contributes to the formation of focal and diffuse pyroglutamate (pGlu)-Abeta deposits in hippocampus via distinct cellular mechanisms. *Acta Neuropathol.* 2011;121:705–19.

[179] Che FY, Lim J, Biswas R, Pan H, Fricker LD. Quantitative neuropeptidomics of microwave-irradiated mouse brain and pituitary. *Mol Cell Proteomics.* 2005;4:1391–405.

[180] Eberwine JH, Barchas JD, Hewlett WA, Evans CJ. Isolation of enzyme cDNA clones by enzyme immunodetection assay: isolation of a peptide acetyltransferase. *Proc Natl Acad Sci USA.* 1987;84:1449–53.

[181] Horvath TL, Diano S, Sotonyi P, Heiman M, Tschop M. Minireview: ghrelin and the regulation of energy balance—a hypothalamic perspective. *Endocrinology.* 2001;142:4163–9.

[182] Yang J, Brown MS, Liang G, Grishin NV, Goldstein JL. Identification of the acyltransferase that octanoylates ghrelin, an appetite-stimulating peptide hormone. *Cell.* 2008;132: 387–96.

[183] Yang HY, Tao T, Iadarola MJ. Modulatory role of neuropeptide FF system in nociception and opiate analgesia. *Neuropeptides.* 2008;42:1–18.

[184] Mzhavia N, Qian Y, Feng Y, Che FY, Devi LA, Fricker LD. Processing of proSAAS in neuroendocrine cell lines. *Biochem J.* 2002;361:67–76.

[185] Mzhavia N, Berman Y, Che FY, Fricker LD, Devi LA. ProSAAS processing in mouse brain and pituitary. *J Biol Chem.* 2001;276:6207–13.

[186] Kaushik S, Arias E, Kwon H, Lopez NM, Athonvarangkul D, Sahu S, Schwartz GJ, Pessin JE, Singh R. Loss of autophagy in hypothalamic POMC neurons impairs lipolysis. *EMBO Rep.* 2012;13:258–65.

[187] Minokadeh A, Funkelstein L, Toneff T, Hwang SR, Beinfeld M, Reinheckel T, Peters C, Zadina J, Hook V. Cathepsin L participates in dynorphin production in brain cortex, illustrated by protease gene knockout and expression. *Mol Cell Neurosci.* 2010;43:98–107.

[188] Beinfeld MC, Funkelstein L, Foulon T, Cadel S, Kitagawa K, Toneff T, Reinheckel T, Peters C, Hook V. Cathepsin L plays a major role in cholecystokinin production in mouse brain cortex and in pituitary AtT-20 cells: protease gene knockout and inhibitor studies. *Peptides.* 2009;30:1882–91.

[189] Funkelstein L, Toneff T, Mosier C, Hwang SR, Beuschlein F, Lichtenauer UD, Reinheckel T, Peters C, Hook V. Major role of cathepsin L for producing the peptide hormones ACTH, beta-endorphin, and alpha-MSH, illustrated by protease gene knockout and expression. *J Biol Chem.* 2008;283:35652–9.

[190] Funkelstein L, Toneff T, Hwang SR, Reinheckel T, Peters C, Hook V. Cathepsin L participates in the production of neuropeptide Y in secretory vesicles, demonstrated by protease gene knockout and expression. *J Neurochem.* 2008;106:384–91.

[191] Hwang SR, Garza C, Mosier C, Toneff T, Wunderlich E, Goldsmith P, Hook V. Cathepsin L expression is directed to secretory vesicles for enkephalin neuropeptide biosynthesis and secretion. *J Biol Chem.* 2007;282:9556–63.

[192] Lim J, Berezniuk I, Che FY, Parikh R, Biswas R, Pan H, Fricker LD. Altered neuropeptide processing in prefrontal cortex of Cpe fat/fat mice: implications for neuropeptide discovery. *J Neurochem.* 2006;96:1169–81.

[193] Goldberg AL. Functions of the proteasome: from protein degradation and immune surveillance to cancer therapy. *Biochem Soc Trans.* 2007;35:12–7.

[194] Fricker LD, Gelman JS, Castro LM, Gozzo FC, Ferro ES. Peptidomic analysis of HEK293T cells: effect of the proteasome inhibitor epoxomicin on intracellular peptides. *J Proteome Res.* 2012;11:1981–90.

[195] Vosler PS, Brennan CS, Chen J. Calpain-mediated signaling mechanisms in neuronal injury and neurodegeneration. *Mol Neurobiol.* 2008;38:78–100.

[196] Pop C, Salvesen GS. Human caspases: activation, specificity, and regulation. *J Biol Chem.* 2009;284:21777–81.

[197] Heitz F, Morris MC, Divita G. Twenty years of cell-penetrating peptides: from molecular mechanisms to therapeutics. *Br J Pharmacol.* 2009;157:195–206.

[198] Stridh MH, Tranberg M, Weber SG, Blomstrand F, Sandberg M. Stimulated efflux of amino acids and glutathione from cultured hippocampal slices by omission of extracellular calcium: likely involvement of connexin hemichannels. *J Biol Chem.* 2008;283:10347–56.

[199] Vaalburg W, Hendrikse NH, Elsinga PH, Bart J, van Waarde A. P-glycoprotein activity and biological response. *Toxicol Appl Pharmacol.* 2005;207:257–60.

[200] Kastin AJ, Fasold MB, Zadina JE. Endomorphins, Met-enkephalin, Tyr-MIF-1, and the P-glycoprotein efflux system. *Drug Metab Dispos.* 2002;30:231–4.

[201] Oude Elferink RP, Zadina J. MDR1 P-glycoprotein transports endogenous opioid peptides. *Peptides.* 2001;22:2015–20.

[202] Nickel W. The unconventional secretory machinery of fibroblast growth factor 2. *Traffic.* 2011;12:799–805.

[203] Holt OJ, Gallo F, Griffiths GM. Regulating secretory lysosomes. *J Biochem.* 2006;140:7–12.

[204] Blott EJ, Griffiths GM. Secretory lysosomes. *Nat Rev Mol Cell Biol.* 2002;3:122–31.

[205] Hegmans JP, Gerber PJ, Lambrecht BN. Exosomes. *Methods Mol Biol.* 2008;484:97–109.

[206] Olver C, Vidal M. Proteomic analysis of secreted exosomes. *Subcell Biochem.* 2007;43: 99–131.

[207] Faure J, Lachenal G, Court M, Hirrlinger J, Chatellard-Causse C, Blot B, Grange J, Schoehn G, Goldberg Y, Boyer V, Kirchhoff F, Raposo G, Garin J, Sadoul R. Exosomes are released by cultured cortical neurones. *Mol Cell Neurosci.* 2006;31:642–8.

[208] Johnstone RM. Exosomes biological significance: a concise review. *Blood Cells Mol Dis.* 2006;36:315–21.

[209] Abele R, Tampe R. The ABCs of immunology: structure and function of TAP, the transporter associated with antigen processing. *Physiology.* 2004;19:216–24.

[210] Kerr MA, Kenny AJ. The purification and specificity of a neutral endopeptidase from rabbit kidney brush border. *Biochem J.* 1974;137:477–88.

[211] Kerr MA, Kenny AJ. The molecular weight and properties of a neutral metallo-endopeptidase from rabbit kidney brush border. *Biochem J.* 1974;137:489–95.

[212] Cudic M, Fields GB. Extracellular proteases as targets for drug development. *Curr Protein Pept Sci.* 2009;10:297–307.

[213] Turner AJ, Nalivaeva NN. New insights into the roles of metalloproteinases in neurodegeneration and neuroprotection. *Int Rev Neurobiol.* 2007;82:113–35.

[214] Roques BP, Fournie-Zaluski MC, Soroca E, Lecomte JM, Malfroy B, Llorens C, Schwartz JC. The enkephalinase inhibitor thiorphan shows antinociceptive activity in mice. *Nature.* 1980;288:286–8.

[215] Matheson AJ, Noble S. Racecadotril. *Drugs.* 2000;59:829–35; discussion 36–7.

[216] Ahn K. Endothelin-converting enzyme 1. In: Barrett AJ, Rawlings ND, Woessner JF, editors. *Handbook of Proteolytic Enzymes.* San Diego, CA: Academic Press; 1998. pp. 1085–9.

[217] Ahn K. Endothelin-converting enzyme 2. In: Barrett AJ, Rawlings ND, Woessner JF, editors. *Handbook of Proteolytic Enzymes.* San Diego, CA: Academic Press; 1998. pp. 1090–1.

[218] Emoto N, Yanagisawa M. Endothelin-converting enzyme-2 is a membrane-bound, phosphoramidon-sensitive metalloprotease with acidic pH optimum. *J Biol Chem.* 1995; 270:15262–8.

[219] Mzhavia N, Pan H, Che FY, Fricker LD, Devi LA. Characterization of endothelin-converting enzyme-2. Implication for a role in the nonclassical processing of regulatory peptides. *J Biol Chem.* 2003;278:14704–11.

[220] Turner AJ, Hooper NM. The angiotensin-converting enzyme gene family: genomics and pharmacology. *Trends Pharmacol Sci.* 2002;23:177–83.

[221] Strittmatter SM, Snyder SH. Angiotensin converting enzyme immunohistochemistry in rat brain and pituitary gland: correlation of isozyme type with cellular localization. *Neuroscience.* 1987;21:407–20.

[222] Warner FJ, Smith AI, Hooper NM, Turner AJ. Angiotensin-converting enzyme-2: a molecular and cellular perspective. *Cell Mol Life Sci.* 2004;61:2704–13.

[223] Turner AJ, Tipnis SR, Guy JL, Rice G, Hooper NM. ACEH/ACE2 is a novel mammalian metallocarboxypeptidase and a homologue of angiotensin-converting enzyme insensitive to ACE inhibitors. *Can J Physiol Pharmacol.* 2002;80:346–53.

[224] Clarke NE, Turner AJ. Angiotensin-converting enzyme 2: the first decade. *Int J Hypertens.* 2012;2012:307315.

[225] Fontenele-Neto JD, Kalinina E, Feng Y, Fricker LD. Identification and distribution of mouse carboxypeptidase A-6. *Mol Brain Res.* 2005;137:132–42.

[226] Skidgel RA. Carboxypeptidase M. In: Barrett AJ, Rawlings ND, Woessner JF, editors. *Handbook of Proteolytic Enzymes.* San Diego, CA: Academic Press; 2004. pp. 851–4.

[227] Tan F, Chan SJ, Steiner DF, Schilling JW, Skidgel RA. Molecular cloning and sequencing of the cDNA for human membrane-bound carboxypeptidase M. *J Biol Chem.* 1989;264:13165–70.

[228] Fricker LD. Carboxypeptidase Z. In: Barrett AJ, Rawlings ND, Woessner JF, editors. *Handbook of Proteolytic Enzymes.* San Diego, CA: Academic Press; 2004. pp. 844–6.

[229] Song L, Fricker LD. Cloning and expression of human carboxypeptidase Z, a novel metallocarboxypeptidase. *J Biol Chem.* 1997;272:10543–50.

[230] Lyons PJ, Fricker LD. Substrate specificity of human carboxypeptidase A6. *J Biol Chem.* 2010;285:38234–42.

[231] Skidgel RA, Davis RM, Tan F. Human carboxypeptidase M: Purification and characterization of a membrane-bound carboxypeptidase that cleaves peptide hormones. *J Biol Chem.* 1989;264:2236–41.

[232] Lyons PJ, Callaway MB, Fricker LD. Characterization of carboxypeptidase A6, an extracellular-matrix peptidase. *J Biol Chem.* 2008;283:7054–63.

[233] Novikova EG, Reznik SE, Varlamov O, Fricker LD. Carboxypeptidase Z is present in the regulated secretory pathway and extracellular matrix in cultured cells and in human tissues. *J Biol Chem.* 2000;275:4865–70.

[234] Novikova EG, Fricker LD, Reznik SE. Carboxypeptidase Z is dynamically expressed in mouse development. *Mech Dev.* 2001;102:259–62.

[235] Reznik SE, Fricker LD. Carboxypeptidases from A to Z: implications in embryonic development and Wnt binding. *Cell Mol Life Sci.* 2001;58:1790–804.

[236] Wang L, Shao YY, Ballock RT. Carboxypeptidase Z (CPZ) links thyroid hormone and Wnt signaling pathways in growth plate chondrocytes. *J Bone Miner Res.* 2009;24: 265–73.

[237] Moeller C, Swindell EC, Kispert A, Eichele G. Carboxypeptidase Z (CPZ) modulates

Wnt signaling and regulates the development of skeletal elements in the chicken. *Development.* 2003;130:5103–11.

[238] Salzmann A, Guipponi M, Lyons PJ, Fricker LD, Sapio M, Lambercy C, Buresi C, Bencheikh BOA, Lahjouji F, Ouazzani R, Crespel A, Chaigne D, Malafosse A. Carboxypeptidase A6 gene (CPA6) mutations in a recessive familial form of febrile seizures and temporal lobe epilepsy and in sporadic temporal lobe epilepsy. *Hum Mutat.* 2012;33:124–35.

[239] Albiston AL, Ye S, Chai SY. Membrane bound members of the M1 family: more than aminopeptidases. *Protein Pept Lett.* 2004;11:491–500.

[240] Turner AJ. Membrane alanyl aminopeptidases. In: Barrett AJ, Rawlings ND, Woessner JF, editors. *Handbook of Proteolytic Enzymes.* San Diego, CA: Academic Press; 1998. pp. 996–1000.

[241] Delmas B, Gelfi J, L'Haridon R, Vogel LK, Sjostrom H, Noren O, Laude H. Aminopeptidase N is a major receptor for the enteropathogenic coronavirus TGEV. *Nature.* 1992; 357:417–20.

[242] Yeager CL, Ashmun RA, Williams RK, Cardellichio CB, Shapiro LH, Look AT, Holmes KV. Human aminopeptidase N is a receptor for human coronavirus 229E. *Nature.* 1992; 357:420–2.

[243] Keller SR. Role of the insulin-regulated aminopeptidase IRAP in insulin action and diabetes. *Biol Pharm Bull.* 2004;27:761–4.

[244] Pessin JE, Thurmond DC, Elmendorf JS, Coker KJ, Okada S. Molecular basis of insulin-stimulated GLUT4 vesicle trafficking. Location! Location! Location! *J Biol Chem.* 1999; 274:2593–6.

[245] Wallis MG, Lankford MF, Keller SR. Vasopressin is a physiological substrate for the insulin-regulated aminopeptidase IRAP. *Am J Physiol Endocrinol Metab.* 2007;293:E1092–102.

[246] Wright JW, Miller-Wing AV, Shaffer MJ, Higginson C, Wright DE, Hanesworth JM, Harding JW. Angiotensin II(3–8) (ANG IV) hippocampal binding: potential role in the facilitation of memory. *Brain Res Bull.* 1993;32:497–502.

[247] Demaegdt H, Lukaszuk A, De Buyser E, De Backer JP, Szemenyei E, Toth G, Chakravarthy S, Panicker M, Michotte Y, Tourwe D, Vauquelin G. Selective labeling of IRAP by the tritiated AT(4) receptor ligand [3H]Angiotensin IV and its stable analog [3H]AL-11. *Mol Cell Endocrinol.* 2009;311:77–86.

[248] Vargas MA, Bourdais J, Sanchez S, Uriostegui B, Moreno E, Joseph-Bravo P, Charli JL. Multiple hypothalamic factors regulate pyroglutamyl peptidase II in cultures of adenohypophyseal cells: role of the cAMP pathway. *J Neuroendocrinol.* 1998;10:199–206.

[249] Cummins PM, O'Connor B. Pyroglutamyl peptidase: an overview of the three known enzymatic forms. *Biochim Biophys Acta.* 1998;1429:1–17.

[250] Charli JL, Vargas MA, Cisneros M, de Gortari P, Baeza MA, Jasso P, Bourdais J, Perez

L, Uribe RM, Joseph-Bravo P. TRH inactivation in the extracellular compartment: role of pyroglutamyl peptidase II. *Neurobiology*. 1998;6:45–57.

[251] Lone AM, Nolte WM, Tinoco AD, Saghatelian A. Peptidomics of the prolyl peptidases. *AAPSJ*. 2010.

[252] Nagase H, Woessner JF. Matrix metalloproteinases. *J Biol Chem*. 1999;274:21491–4.

[253] Clements JA. Reflections on the tissue kallikrein and kallikrein-related peptidase family—from mice to men—what have we learnt in the last two decades? *Biol Chem*. 2008;389: 1447–54.

[254] Yang P, Baker KA, Hagg T. The ADAMs family: coordinators of nervous system development, plasticity and repair. *Prog Neurobiol*. 2006;79:73–94.

[255] Allinson TM, Parkin ET, Turner AJ, Hooper NM. ADAMs family members as amyloid precursor protein alpha-secretases. *J Neurosci Res*. 2003;74:342–52.

[256] Pierotti A, Dong KW, Glucksman MJ, Orlowski M, Roberts JL. Molecular cloning and primary structure of rat testes metalloendopeptidase EC 3.4.24.15. *Biochemistry*. 1990;29:10323–9.

[257] Ferro ES, Carreno FR, Goni C, Garrido PA, Guimaraes AO, Castro LM, Oliveira V, Araujo MC, Rioli V, Gomes MD, Fontenele-Neto JD, Hyslop S. The intracellular distribution and secretion of endopeptidases 24.15 (EC 3.4.24.15) and 24.16 (EC 3.4.24.16). *Protein Pept Lett*. 2004;11:415–21.

[258] Russo LC, Goni CN, Castro LM, Asega AF, Camargo AC, Trujillo CA, Ulrich H, Glucksman MJ, Scavone C, Ferro ES. Interaction with calmodulin is important for the secretion of thimet oligopeptidase following stimulation. *FEBS J*. 2009;276:4358–71.

[259] Bulloj A, Leal MC, Xu H, Castano EM, Morelli L. Insulin-degrading enzyme sorting in exosomes: a secretory pathway for a key brain amyloid-beta degrading protease. *J Alzheimers Dis*. 2010;19:79–95.

[260] Jordans S, Jenko-Kokalj S, Kuhl NM, Tedelind S, Sendt W, Bromme D, Turk D, Brix K. Monitoring compartment-specific substrate cleavage by cathepsins B, K, L, and S at physiological pH and redox conditions. *BMC Biochem*. 2009;10:23.

[261] Myers RD. Neuroactive peptides: unique phases in research on mammalian brain over three decades. *Peptides*. 1994;15:367–81.

[262] Jiang N, Kolhekar AS, Jacobs PS, Mains RE, Eipper BA, Taghert PH. PHM is required for normal developmental transitions and for biosynthesis of secretory peptides in Drosophila. *Dev Biol*. 2000;226:118–36.

[263] Morgan DJ, Wei S, Gomes I, Czyzyk T, Mzhavia N, Pan H, Devi LA, Fricker LD, Pintar JE. The propeptide precursor proSAAS is involved in fetal neuropeptide processing and body weight regulation. *J Neurochem*. 2010;113:1275–84.

[264] Chakraborty TR, Tkalych O, Nanno D, Garcia AL, Devi LA, Salton SR. Quantification of VGF- and pro-SAAS-derived peptides in endocrine tissues and the brain, and their regulation by diet and cold stress. *Brain Res.* 2006;1089:21–32.

[265] Lindberg I, Yang HY. Distribution of Met5-enkephalin-Arg6-Gly7-Leu8-immunoreactive peptides in rat brain: presence of multiple molecular forms. *Brain Res.* 1984;299:73–8.

[266] Zhao E, Zhang D, Basak A, Trudeau VL. New insights into granin-derived peptides: evolution and endocrine roles. *Gen Comp Endocrinol.* 2009;164:161–74.

[267] Nielsen E, Welinder BS, Madsen OD. Chromogranin-B, a putative precursor of eight novel rat glucagonoma peptides through processing at mono-, di-, or tribasic residues. *Endocrinology.* 1991;129:3147–56.

[268] Martens GJ, Braks JA, Eib DW, Zhou Y, Lindberg I. The neuroendocrine polypeptide 7B2 is an endogenous inhibitor of prohormone convertase PC2. *Proc Natl Acad Sci USA.* 1994;91:5784–7.

[269] Matsuda K, Yuzaki M. Cbln family proteins promote synapse formation by regulating distinct neurexin signaling pathways in various brain regions. *Eur J Neurosci.* 2011;33:1447–61.

[270] Wu G, Feder A, Wegener G, Bailey C, Saxena S, Charney D, Mathe AA. Central functions of neuropeptide Y in mood and anxiety disorders. *Exp Opin Ther Targets.* 2011;15:1317–31.

[271] Zhang L, Bijker MS, Herzog H. The neuropeptide Y system: pathophysiological and therapeutic implications in obesity and cancer. *Pharmacol Ther.* 2011;131:91–113.

[272] Kakidani H, Furutani Y, Takahashi H, Noda M, Morimoto Y, Hirose T, Asai M, Inayama S, Nakanishi S, Numa S. Cloning and sequence analysis of cDNA for porcine beta-neo-endorphin/dynorphin precursor. *Nature.* 1982;298:245–9.

[273] Noda M, Furutani Y, Takahashi H, Toyosato M, Hirose T, Inayama S, Nakanishi S, Numa S. Cloning and sequence analysis of cDNA for bovine adrenal preproenkephalin. *Nature.* 1982;295:202–6.

[274] Roberts JL, Seeburg PH, Shine J, Herbert E, Baxter JD, Goodman HM. Corticotropin and beta-endorphin: construction and analysis of recombinant DNA complementary to mRNA for the common precursor. *Proc Natl Acad Sci USA.* 1979;76:2153–7.

[275] Akil H, Owens C, Gutstein H, Taylor L, Curran E, Watson S. Endogenous opioids: overview and current issues. *Drug Alcohol Depend.* 1998;51:127–40.

[276] Feng Y, Reznik SE, Fricker LD. Distribution of proSAAS-derived peptides in rat neuroendocrine tissues. *Neuroscience.* 2001;105:469–78.

[277] Cameron A, Fortenberry Y, Lindberg I. The SAAS granin exhibits structural and functional homology to 7B2 and contains a highly potent hexapeptide inhibitor of PC1. *FEBS Lett.* 2000;473:135–8.

[278] Qian Y, Devi LA, Mzhavia N, Munzer S, Seidah NG, Fricker LD. The C-terminal region of proSAAS is a potent inhibitor of prohormone convertase 1. *J Biol Chem.* 2000;275: 23596–601.

[279] Wardman JH, Berezniuk I, Di S, Tasker JG, Fricker LD. ProSAAS-derived peptides are colocalized with neuropeptide Y and function as neuropeptides in the regulation of food intake. *PloS One.* 2011;6:e28152.

[280] Wei S, Feng Y, Che FY, Pan H, Mzhavia N, Devi L, McKenzie AA, Levin N, Richards WG, Fricker LD. Obesity and diabetes in transgenic mice expressing proSAAS. *J Endocrinol.* 2004;180 357–68.

[281] Thanawala V, Kadam VJ, Ghosh R. Enkephalinase inhibitors: potential agents for the management of pain. *Curr Drug Targets.* 2008;9:887–94.

[282] Wisner A, Dufour E, Messaoudi M, Nejdi A, Marcel A, Ungeheuer MN, Rougeot C. Human Opiorphin, a natural antinociceptive modulator of opioid-dependent pathways. *Proc Natl Acad Sci USA.* 2006;103:17979–84.

[283] Guidotti A, Forchetti CM, Corda MG, Konkel D, Bennett CD, Costa E. Isolation, characterization, and purification to homogeneity of an endogenous polypeptide with agonistic action on benzodiazepine receptors. *Proc Natl Acad Sci USA.* 1983;80:3531–5.

[284] Costa E, Guidotti A. Diazepam binding inhibitor (DBI): a peptide with multiple biological actions. *Life Sci.* 1991;49:325–44.

[285] Biagioli M, Pinto M, Cesselli D, Zaninello M, Lazarevic D, et al. Unexpected expression of alpha- and beta-globin in mesencephalic dopaminergic neurons and glial cells. *Proc Natl Acad Sci USA.* 2009;106:15454–9.

[286] Richter F, Meurers BH, Zhu C, Medvedeva VP, Chesselet MF. Neurons express hemoglobin alpha and beta chains in rat and human brains. *J Comp Neurol.* 2009;51 538–47.

[287] Liu L, Zeng M, Stamler JS. Hemoglobin induction in mouse macrophages. *Proc Natl Acad Sci USA.* 1999;96:6643–7.

[288] Lin Y, Hall RA, Kuhar MJ. CART peptide stimulation of G protein-mediated signaling in differentiated PC12 cells: identification of PACAP 6–38 as a CART receptor antagonist. *Neuropeptides.* 2011;45:351–8.

[289] Vicentic A, Lakatos A, Jones D. The CART receptors: background and recent advances. *Peptides.* 2006;27:1934–7.

[290] Lembo PMC, Grazzini E, Groblewski T, O'Donnell D, Roy M, et al. Proenkephalin A gene products activate a new family of sensory neuron-specific GPCRs. *Nat Neurosci.* 2002;5:201–9.

[291] Breit A, Gagnidze K, Devi LA, Lagace M, Bouvier M. Simultaneous activation of the delta opioid receptor (deltaOR)/sensory neuron-specific receptor-4 (SNSR-4) hetero-oligomer

by the mixed bivalent agonist bovine adrenal medulla peptide 22 activates SNSR-4 but inhibits deltaOR signaling. *Mol Pharmacol.* 2006;70:686–96.

[292] Drake MT, Violin JD, Whalen EJ, Wisler JW, Shenoy SK, Lefkowitz RJ. Beta-arrestin-biased agonism at the beta2-adrenergic receptor. *J Biol Chem.* 2008;283:5669–76.

[293] Berezikov E. Evolution of microRNA diversity and regulation in animals. *Nat Rev Genet.* 2011;12:846–60.

[294] Haynes CM, Yang Y, Blais SP, Neubert TA, Ron D. The matrix peptide exporter HAF-1 signals a mitochondrial UPR by activating the transcription factor ZC376.7 in C. elegans. *Mol Cell.* 2010;37:529–40.

[295] Kondo T, Plaza S, Zanet J, Benrabah E, Valenti P, Hashimoto Y, Kobayashi S, Payre F, Kageyama Y. Small peptides switch the transcriptional activity of Shavenbaby during Drosophila embryogenesis. *Science.* 2010;329:336–9.

[296] Tagliabracci VS, Engel JL, Wen J, Wiley SE, Worby CA, Kinch LN, Xiao J, Grishin NV, Dixon JE. Secreted Kinase Phosphorylates Extracellular Proteins That Regulate Biomineralization. *Science.* June 1, 2012;336.